Cafe House 1

Cafe House 1

출판사 : 바른손
저　자 : 김 연 경
발행일 : 2024. 08. 20
주　소 : 경기도 파주시 월롱면 덕은리 216-17
가　격 : ₩98,000원

Coffea Arabica - C. Arabica

Varieties: Bourbon, Typica, Caturra, Mundo Novo, Tico, [S]an Ramon, Jamaican Blue Mountain

[C]offea Arabica is descended from the original coffee [t]rees discovered in Ethiopia. These trees produce a fine, [m]ild, aromatic coffee and represent approximately 70% [o]f the world's coffee production. The beans are flatter [a]nd more elongated than Robusta and lower in caffeine. [O]n the world market, Arabica coffees bring the highest [p]rices. The better Arabicas are high grown coffees — [g]enerally grown between 2,000 to 6,000 feet (610 to [1]830 meters) above sea level - though optimal altitude [v]aries with proximity to the equator. The most import[a]nt factor is that temperatures must remain mild, ideally [b]etween 59 - 75 degrees Fahrenheit, with about 60 [i]nches of rainfall a year. The trees are hearty, but a heavy [f]rost will kill them. Arabica trees are costly to cultivate [b]ecause the ideal terrain tends to be steep and access is [d]ifficult. Also, because the trees are more disease-prone [t]han Robusta, they require additional care and attention.

Coffea canephora - C. canephora var. Robusta

Variety: Robusta

Most of the world's Robusta is grown in Central and Western Africa, parts of Southeast Asia, including Indonesia and Vietnam, and in Brazil. Production of Robusta is increasing, though it accounts for only about 30% of the world market. Robusta is primarily used in blends and for instant coffees. The Robusta bean itself tends to be slightly rounder and smaller than an Arabica bean. The Robusta tree is heartier and more resistant to disease and parasites, which makes it easier and cheaper to cultivate. It also has the advantage of being able to withstand warmer climates, preferring constant temperatures between 75 and 85 degrees Fahrenheit, which enables it to grow at far lower altitudes than Arabica. It requires about 60 inches of rainfall a year, and cannot withstand frost. Compared with Arabica, Robusta beans produce a coffee which has a distinctive taste and about 50-60% more caffeine.

Coffee shops & Cafe design?

Coffee shops & Cafe interior design is..... a choosing furniture and wall colors, though, it's important to layout your coffee shop floor plan. Your floor plan is going to dictate many components of your customer experience–from how they line up, to where they sit, to how long they sit, to what food and drinks you are able to serve, and more.

Here are seven things to think about when designing a coffee shop floor plan.

1. Prioritize a layout that fits your equipment.

It's fun to think about design elements and furnishing for your customers, but the first thing to prioritize is the large equipment you'll need to bring those customers in the door.

We tell our students to finish planning their menu before laying out their coffee shop floor plan. The menu will dictate what kind of equipment you need in your kitchen and on the coffee bar. Start at the back of the house and design forward. This ensures you'll have enough space for your specific equipment and workflow.

Finally, equipment needs to be able to be able to easily come in and out of your coffee shop for installation, service, cleaning, or replacement. When designing your coffee shop layout, be mindful of this important detail. Make sure it can all fit through your doors and around your walls and counters.

2. Make sure your coffee shop floor plan provides employees optimal space to work efficiently.

Finding the right amount of space in your employee work station is a balancing act. To be efficient and safe, you want to avoid making employees cross paths often. You also want to avoid making them run all the way down the bar and back.

If the space is too big, though, they'll have to take a lot of steps to serve customers. Lay out your workspace so employees have stations and can rotate 360 degrees to touch everything they need to do their job on that station. That way, they won't have to pass around each other, and you'll free up space in the front for customers.

3. Use vertical space for storage.

Make sure you have the highest ceilings possible in the back of the house (kitchen and bar area.) Install shelving all the way up for maximum storage capacity for large boxes of paper products, coffee, and more.

If you have a drop ceiling in the kitchen or storage areas tear it out! You want to make sure you can (safely) utilize all of that vertical space for your storage needs.

4. Choose the right furniture.

Your home living room is most likely cozy, but avoid replicating that same atmosphere in your coffee shop design. Having couches significantly limits seating and your ability to serve the maximum number of customers.

Also consider that most strangers rather not sit close together in communal seating areas (especially in the post COVID world.) Often, two people will take up a large eight top table and nobody else will sit at the table with them. This is highly inefficient, and makes for a bad user experience for your other customers looking for a place to sit. We recommend smaller two-top tables so you can maximize on seating for the most customers. You can always push them together to accommodate larger parties, then separate them back out again when they leave.

It's also important to consider how much time your customers plan to stay. Power outlets and large tables make for great work spaces. But, the longer your customers plan to stay, the more you need them to purchase. If you're planning on creating a great place to hang out for hours, make sure your menu accommodates your guests with snacks, lunch, or even alcohol options.

5. Make cleaning, health, and safety easier.

Cleaning is even more important now than ever before. To make cleaning and sanitizing easier, the ceilings and back splash areas in your workspace need to be smooth (nonporous) and wipeable. Make sure the materials can withstand commercial levels of frequent cleaning with strong sanitizing and cleaning agents. In accordance with most city health codes, be sure to use "cove base" (curved) tile/flooring around the perimeter of the bar and kitchen areas where the floor meets the wall.

6. Make sure to install sneeze guards over any pastry displays and possibly between your customers and employees. When possible, many customers are opting for contact-free options to get their coffee.

7. Design your coffee shop floor plan for the post-pandemic world.

It's imperative you take into account the changing ways customers are interacting with coffee shops after COVID-19. During the pandemic, we watched as coffee shop owners found innovative ways to continue serving their customers. Drive thrus, walk-up windows, curbside pick-up, delivery service pick-up areas, and more outdoor seating became imperative.

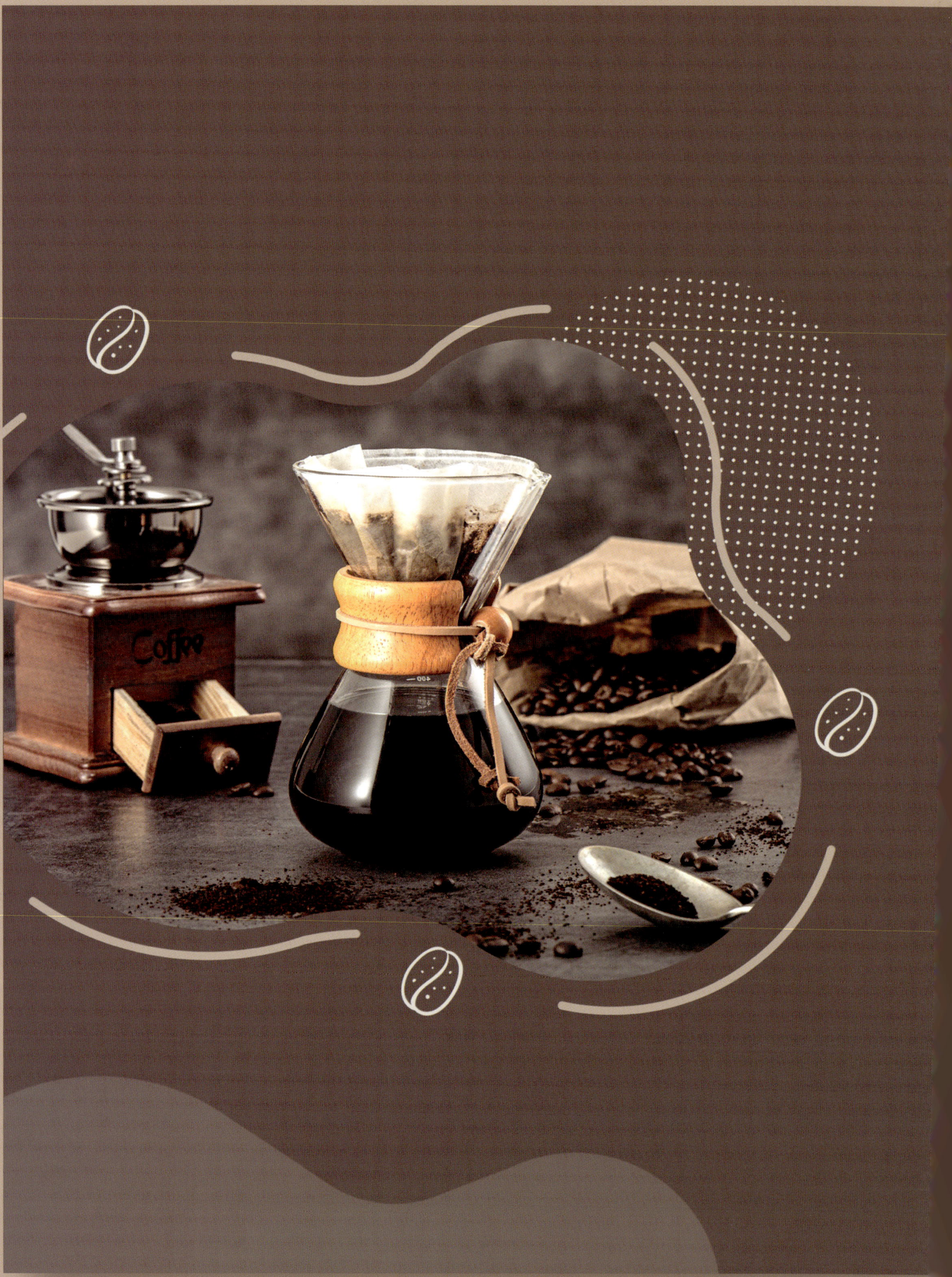

Table of contents

008~019	**Dessert cafe, Dyce**	Formroom
020~031	**Cafe Dubler**	Balbek Bureau
032~041	**RAWR Cafe**	OwT Branding & Design
042~049	**Coperaco Coffee**	Concrete
050~061	**Myfresh Café**	Loop Design Studio
062~077	**Blue Bottle Coffee**	Jo Nagasaka, Schemata Architects
078~099	**Kava & Chai coffee**	4 SPACE Design
100~111	**House of Eden**	Ritz Ghougassian
112~123	**Bogen Bistro & Cafe**	Noa* network of architecture
124~133	**Cafe & Salon Sacher**	BWM Architekten und Partner
134~143	**Café Flor**	Creneau.com Int
144~155	**Barn's Cafe**	ODG designs
156~167	**La Ganea**	Studio Mabb
168~181	**Kaizen coffee co**	Pittawas Euavongkul
182~195	**GAGA King Glory Cafe**	MA design
196~207	**Vòm Coffee**	Emem Design
208~221	**Pano Brot & Kaffee**	DIA-Dittel Architekten
222~235	**NANA Coffee Roasters**	IDIN Architects
236~245	**Primo Cafébar**	DIA - Dittel Architekten
246~253	**Cupping Room**	M.r. Studio
254~269	**September Coffee Shop**	Red5studio + Ben Decor
270~281	**Melrose Coffee**	House of Forme
282~289	**Abbocca Bistro Cafe**	Ramoprimo Architects
290~303	**YAMA Coffee Shop**	Ksoul Studio

Dessert cafe, Dyce
London, United Kingdom

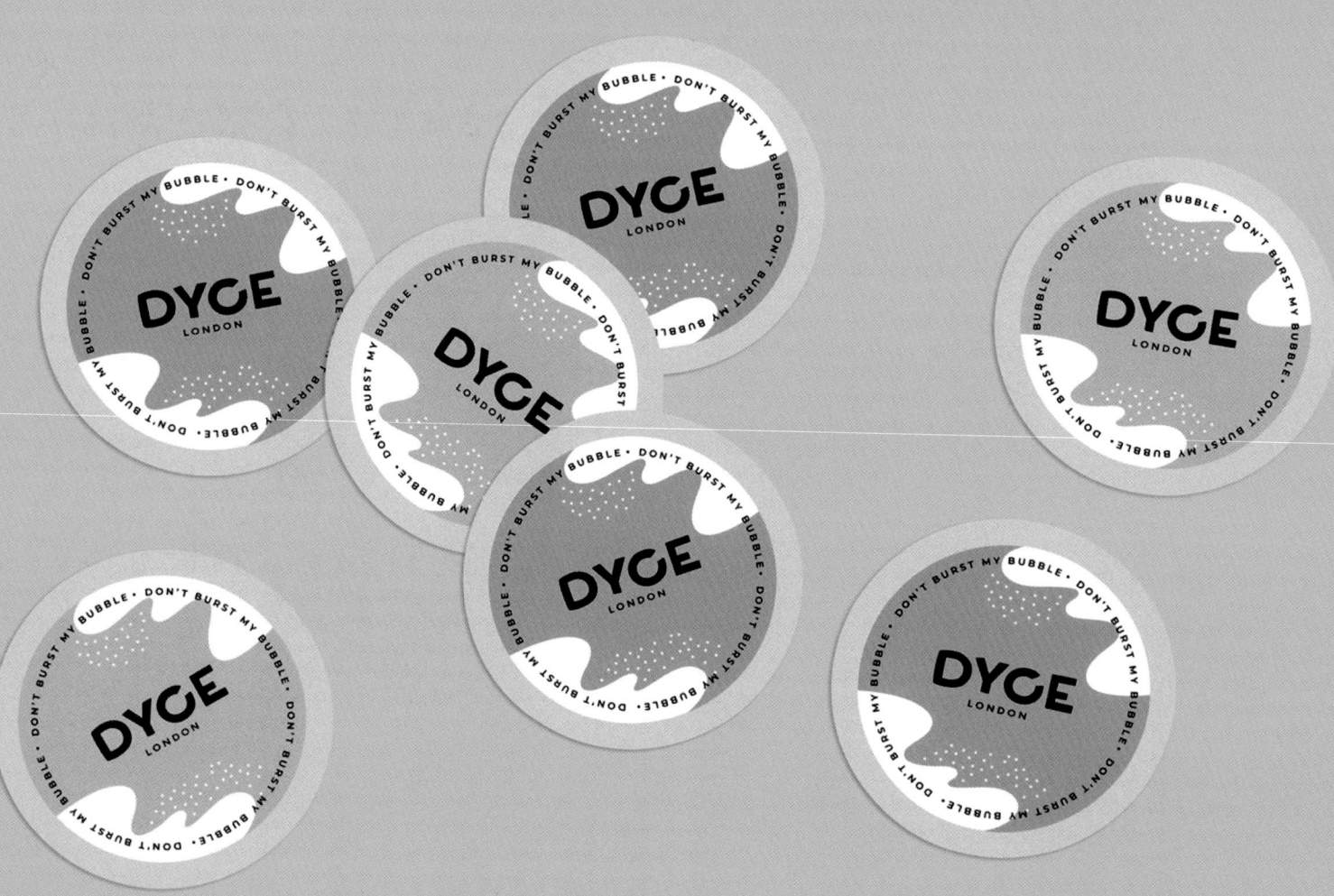

Formroom?

Formroom is a global leader of design, collaborating with the best brands and future stars of the retail, hospitality and commercial industries. For our clients, we want to be their trusted brand guardian across all touch points, specifically within their brand identity and spaces.
We're a creatively ambitious team of brand strategists, designers and project managers from a broad range of backgrounds, skill sets and points of view. We want to inspire and support each other to create incredible work and show passion whilst we're doing it.

Design: *formroom*   **Homepage:** *formroom.com/projects/dyce*   **Team:** *Nina Rowland, Emily Ditton, Alexandra Motovilina......*
Location: *Marylebone, London, United Kingdom*   **Photographs:** *Marcus Peel*

We blended playfulness and surrealism with a contemporary edge for emerging dessert parlour, Dyce.

Whilst the brief called for an 'Instagramable' interior, FormRoom pushed to develop the surreal café design concept even further, incorporating elements that invite playful interaction in movement to also appeal to video sharing platforms like Tik Tok or Snapchat. Evolving upon the three brand pillars; Approachable, Unexpected and Immersive, FormRoom created a memorable aesthetic with a fusion of youthfulness and Salvador Dali inspired surrealism. The space illustrates a conceptual take on the desert ingredients, including melting ice cream and bubble tea. A moment of childlike escapism balanced with a contemporary edge. The bustling hub of James St, Marylebone provides the design blueprint and first permanent space for the Dyce brand.

Insight & Strategy

It was important to consider how the store could feel approachable to all ages while catering to the different needs of social sharing demographics. While the bubble feature chairs appeal to the Instagram driven millennials, Gen Z look for complete store experiences and surreal café design where they can capture and share the whole journey through video – a more authentic representation of their experience. The amphitheatre-style seating allowed the space to feel welcoming to families while limiting excessive dwell time in favour of a faster customer journey and experience.

Design & Manufacture

The colour palette expands on the pink theme, but with unexpected but complimentary colour contrasts. The bold shapes, textures and pastel coloured interior ignite customer curiosity while catering to the demand for shareable moments.

The curvaceous two-tiered seating is inspired by the smooth dripping nature of soft serve ice cream. Bespoke upholstered cocoon seating sits parallel, nestled within a field of soft pastel bubbles and surrounded in a warm neon glow. Above, the brand signage is accented by LED trimming and set against a pastel pink backdrop.

Dessert visual cues continue throughout the space with curved floor artwork to mimic the ripples of ice cream; an unexpected wayfinding and subtle nod to the surrealist aesthetic. The use of a corrugated metal panels treated with an iridescence vinyl continue the juxtaposition of playful and rustic edge to the counter.

The striking ceiling installation combines suspended concave and convex iridescent mirrors, representing bubbles from Dyce's core product offering. Subtle accents of black trimming give the soft colour palette a defined finish.

The mirror wall acts as an extension to the ceiling installation, bordering a large convex mirror encircled with the brand phrase 'Don't Burst My Bubble.' A unique 'fish-eye' effect and light-enhancing feature. The distorted visuals reflect the interior space and cater to the young demographic drawn to spaces with shareable moments.

"Dyce is a place where people can escape the real world and enjoy a moment of joy away from the everyday"

Zahra Khan, founder of Dyce

With its insta-worthy treats and melting ice cream theme, DYCE brings a modern and electric vibe to the heart of London. This new cute and eye-popping cafe offers various exhilarating bubble tea and gelato flavours with vegan and gluten-free options paired with a delicious mash-up of toppings to make your dreamed-treat come true. The eye-catching and trendy décor makes you feel like you are encapsulated in melting ice cream.

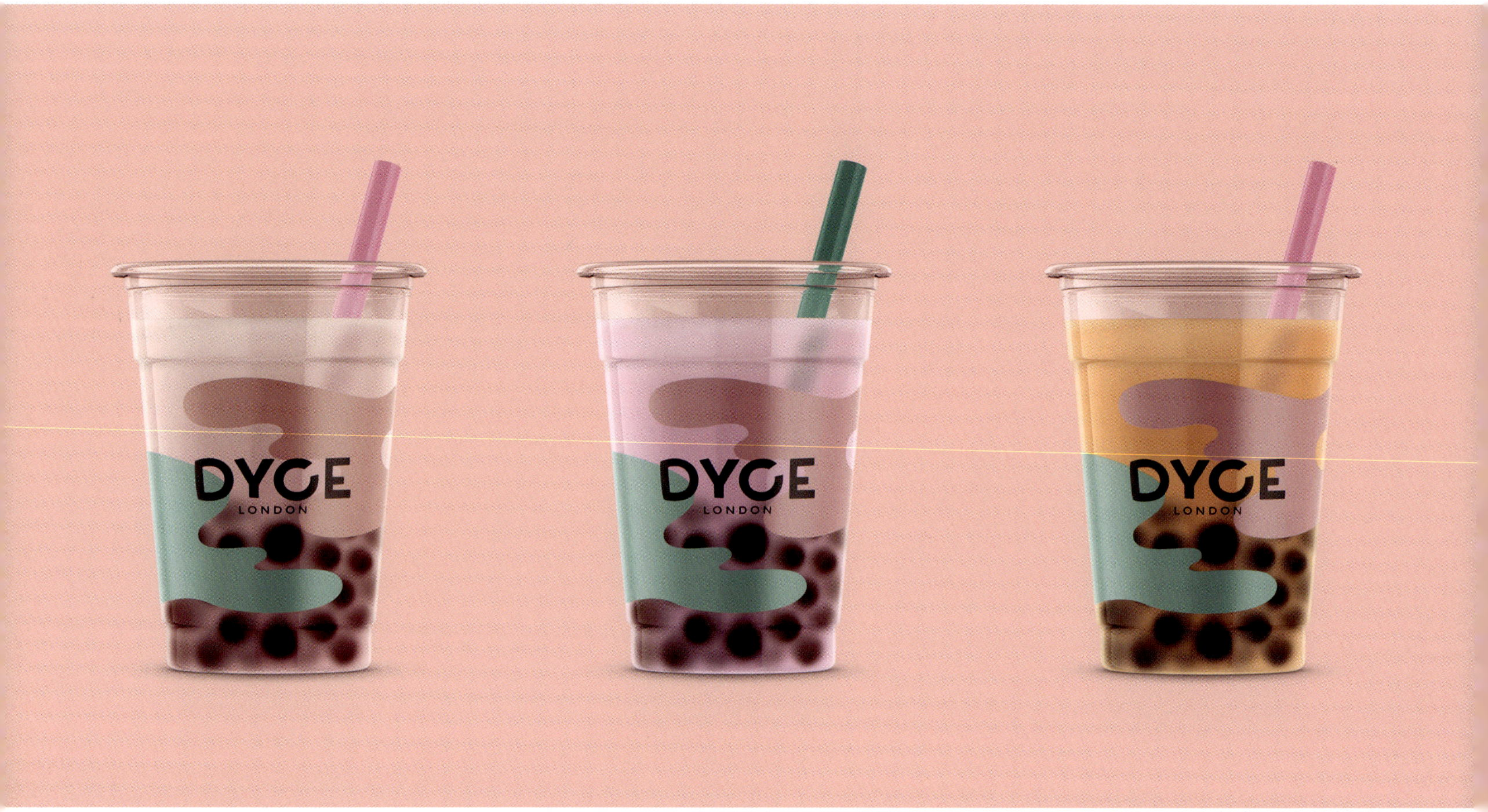
DYCE
LONDON
DYCE
LONDON
DYCE
LONDON

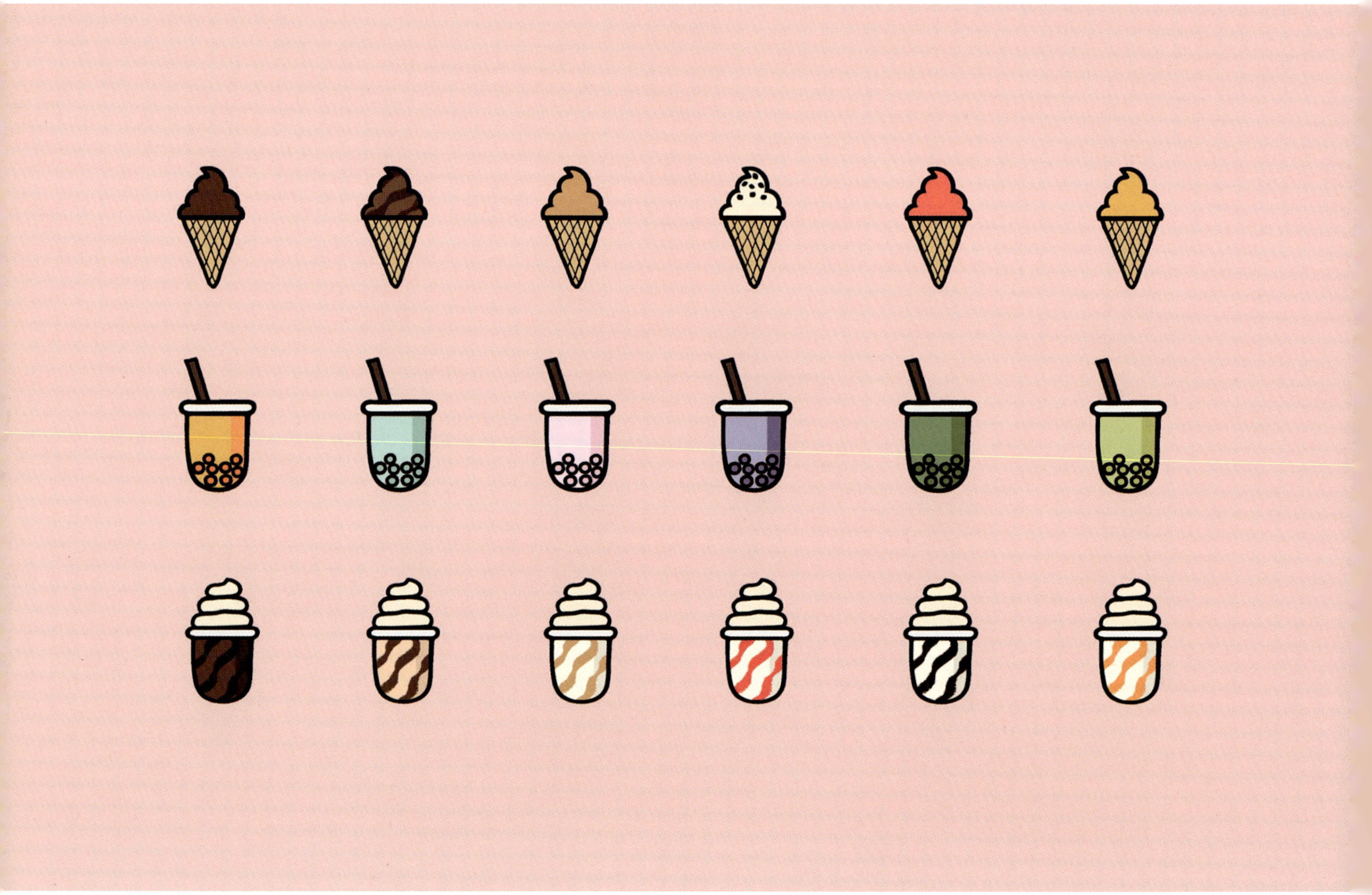

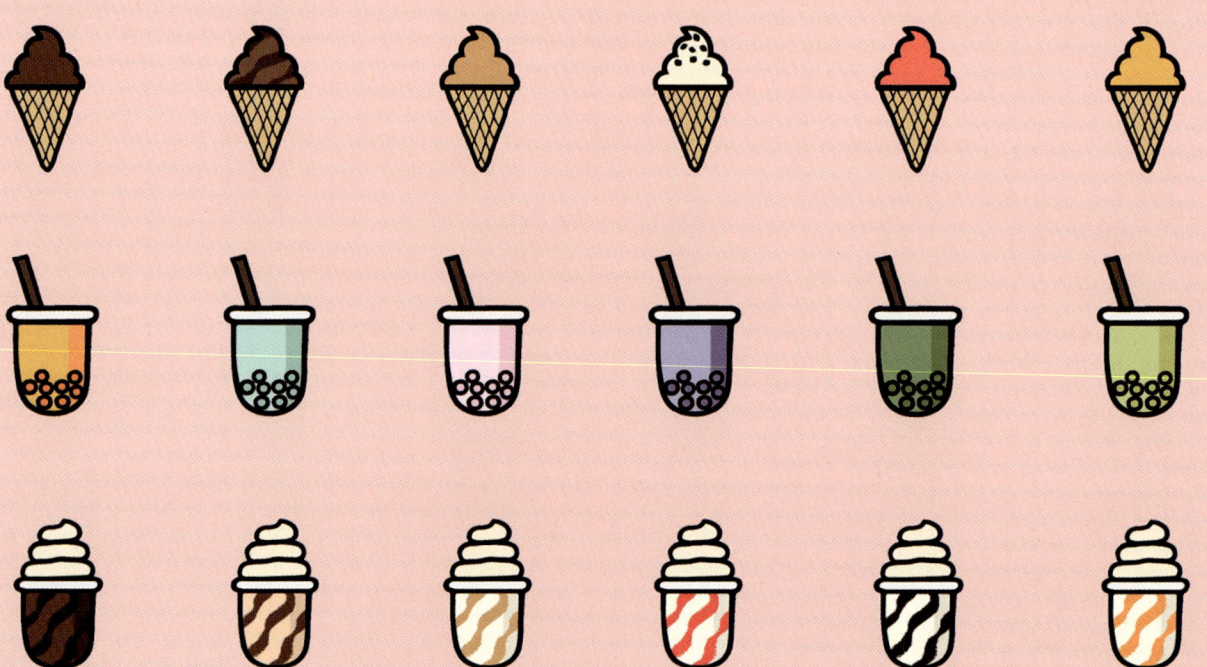

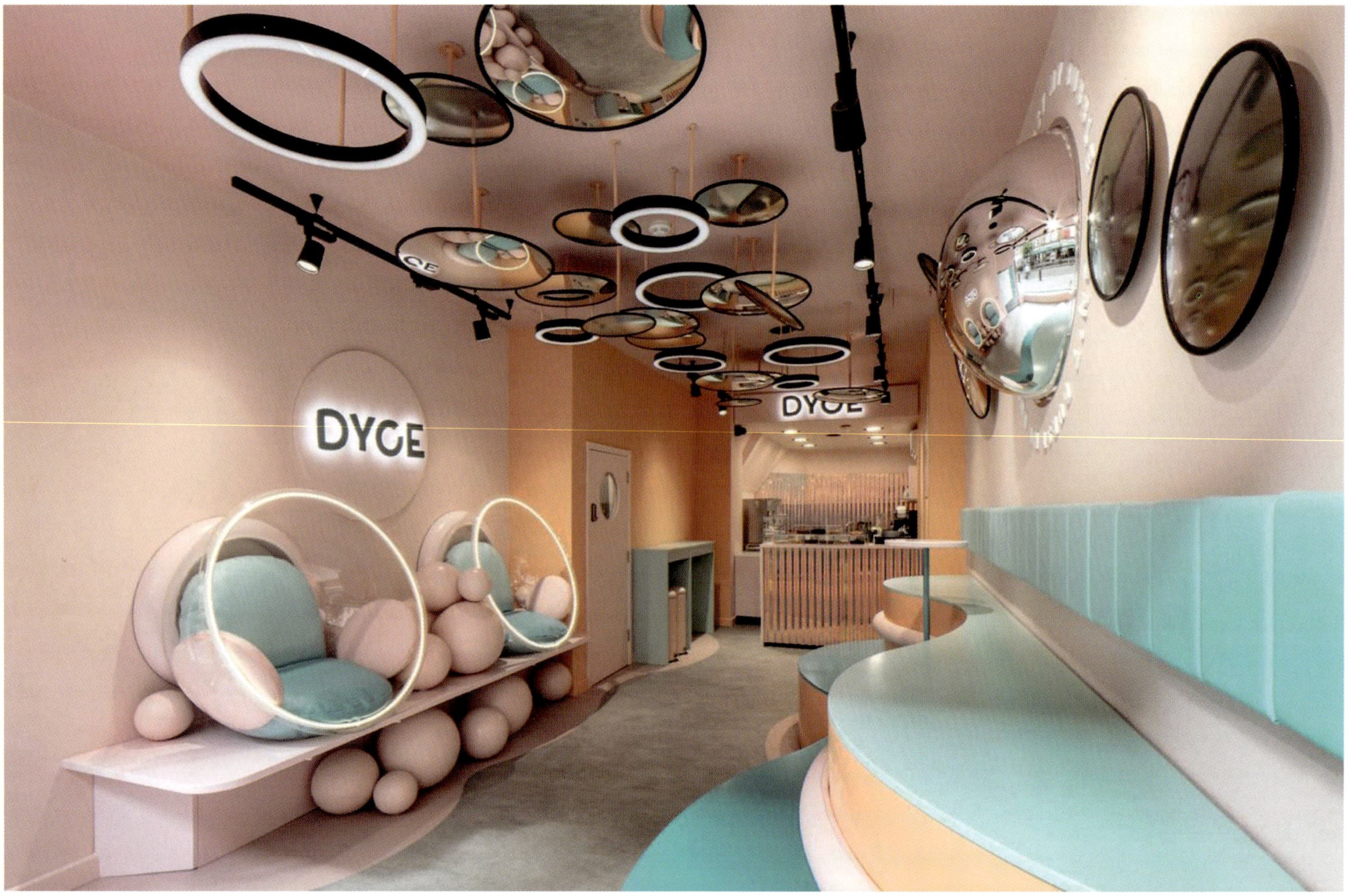
DYCE
DYCE

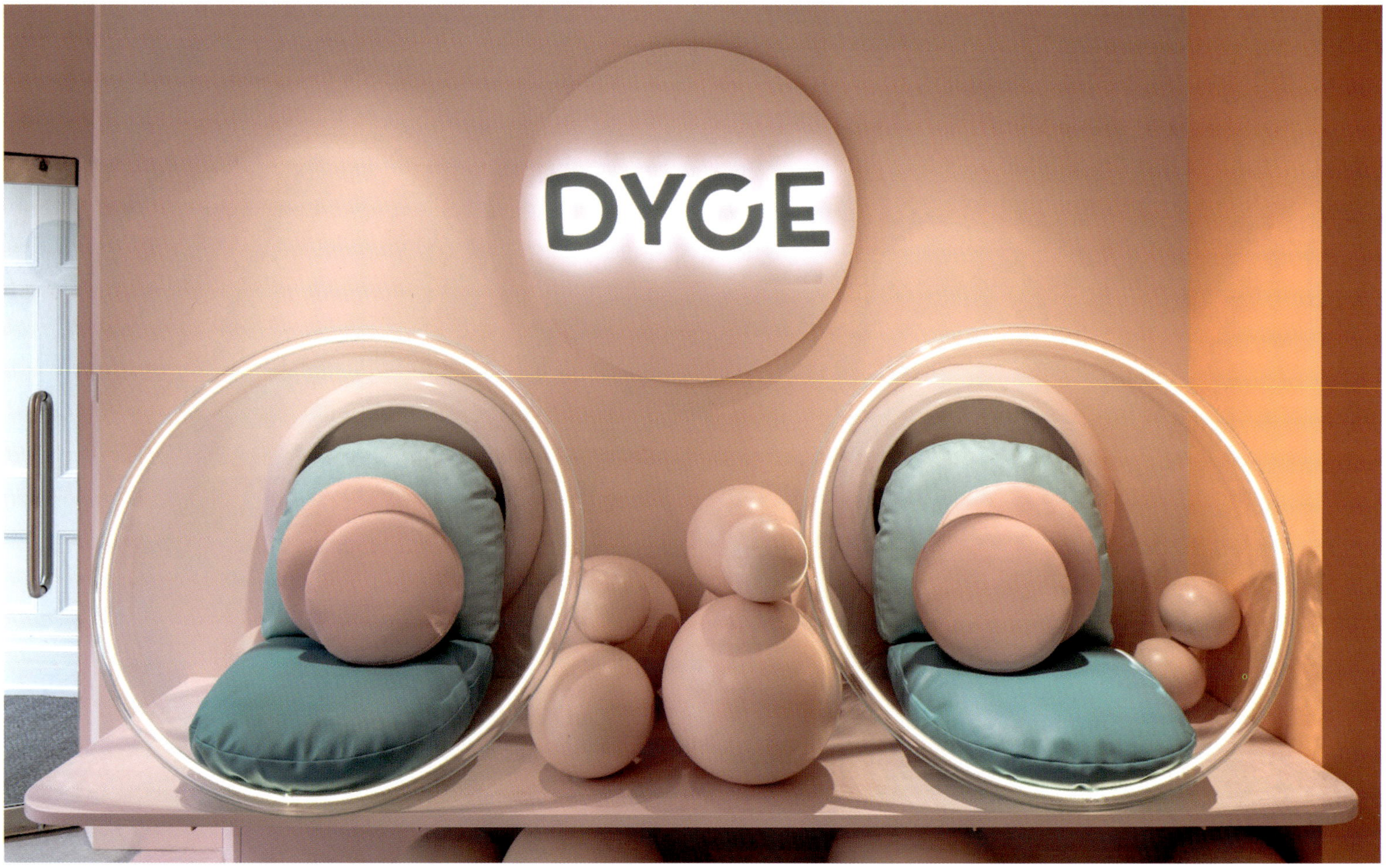
DYCE

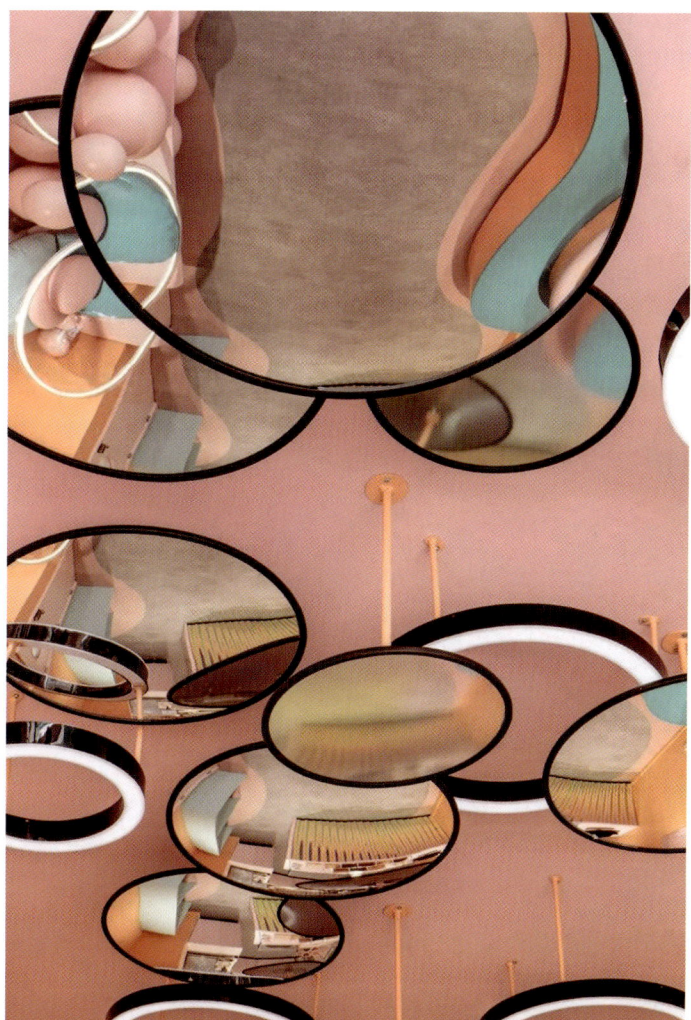

018

Cafe Dubler
Kyiv, Ukraine

дублер

Balbek Bureau?

Balbek Bureau is an award-winning architecture and interior design studio founded by Ukrainian architect Slava Balbek and Borys Dorogov. For 14 years, we have been designing bespoke commercial, corporate and residential spaces. Comfort, innovation and functionality are the driving forces behind every project we work on. Our approach is to explore the basics and then plunge into details to transform aspirations into ground-breaking environments. Our work has received multiple international awards and has been published in numerous media outlets worldwide.

Design: *Balbek Bureau*   **Homepage:** *www.balbek.com*   **Team:** *Slava Balbek, Anastasiia Partyka, Alina Vovkotrub, Mariia Kovalenko*
Location: *Pechersk district of Kyiv, Ukraine*   **Project area:** *168 sq.m*   **Photographs:** *Yevhenii Avramenko, Maryan Beresh*

Dubler is a city cafe in the Pechersk district of Kyiv. The co-owners of the cafe are the founder of balbek bureau Slava Balbek and hotelier Roman Tatarsky, who launched Dyletant cafe last year, as well as CFO and COO balbek bureau Julia Kolesnikova and Boris Dorogov, along with the head of Dyletant Sashko Borovsky.

"We wanted to move the Podil vibe to Pechersk, so we deliberately looked for premises in old and not new buildings," says Slava Balbek. The perfect place was spotted on John Paul II Street, on the ground floor of a residential building erected in the 70s. The founder calls this location the 'eye of the storm', alluding to how the high-rise residential estates tower above the old building. The café with a terrace and balcony is located in a cozy courtyard. The facade of the building with arched windows is partially covered with grapevines.

Concept

The main value of the Dubler is the community that is formed around its cool cuisine and easy-going service. With this in mind, we decided to create a cozy interior that will allow guests to simultaneously enjoy the food and socialize. The architecture and atmosphere of the old building were emphasized with the help of vintage furniture and decor. Dubler's vibe resembles a flea market of the '70s, but due to its simple and pure forms, it does not lose touch with the modern world.

Construction

Previously on the site of the cafe stood an office and an apartment. To combine these rooms into one single space, we dismantled several partitions. The openings were reinforced and lined with plasterboard. The floor was unfit for further use, so we deconstructed it and made new, self-levelling flooring. The ventilation and electrical networks were fixed on the ceiling and left exposed. The walls were covered with several layers of transparent primer. There was no additional finishing work to the interior. The bathroom is constructed out of solid aerated concrete blocks. In order for the wall leading to the main hall to be of the needed length, we had to trim each element symmetrically. This was challenging because aerated concrete blocks are quite fragile.

The main hall is located right by the entrance. The first thing the guests of Dubler see is the bar and pastry counter, with friendly baristas at work. A communal table extends parallel to the bar - probably the largest designed by our team. Soft seating is located to the right of the entrance, behind which begins a small hall. With traditional dining-style seating, the hall seats two or four people per table. The area is visually enlarged by a large mirror on one of the walls. The far part of the hall was set aside for technical rooms and a kitchen, which, at the request of the chef, was to be spacious and comfortable. Bathrooms were set up nearby, and an area where you can wash your hands was placed by the grapevine-adorned arched windows.

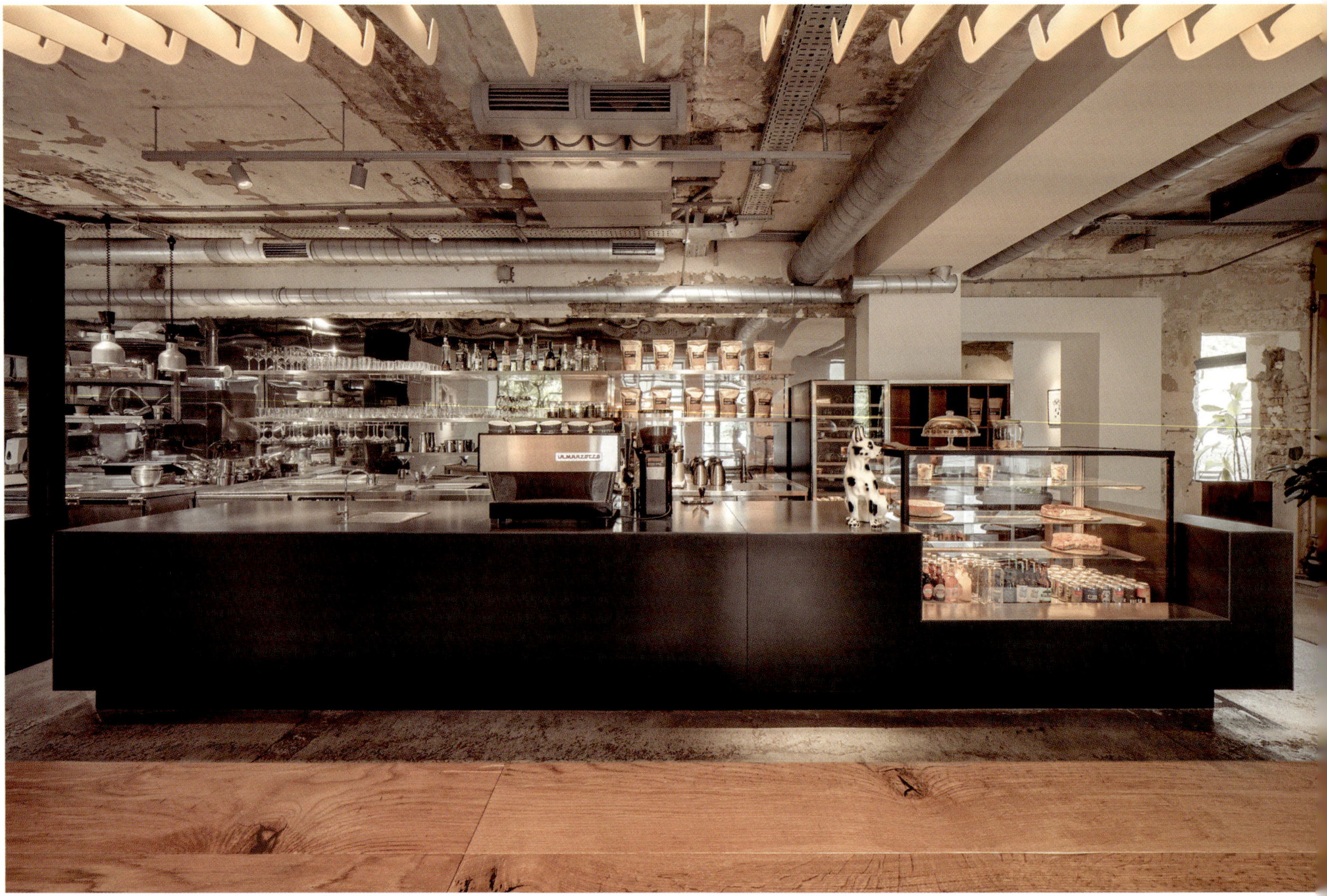
LA MARZOCCO

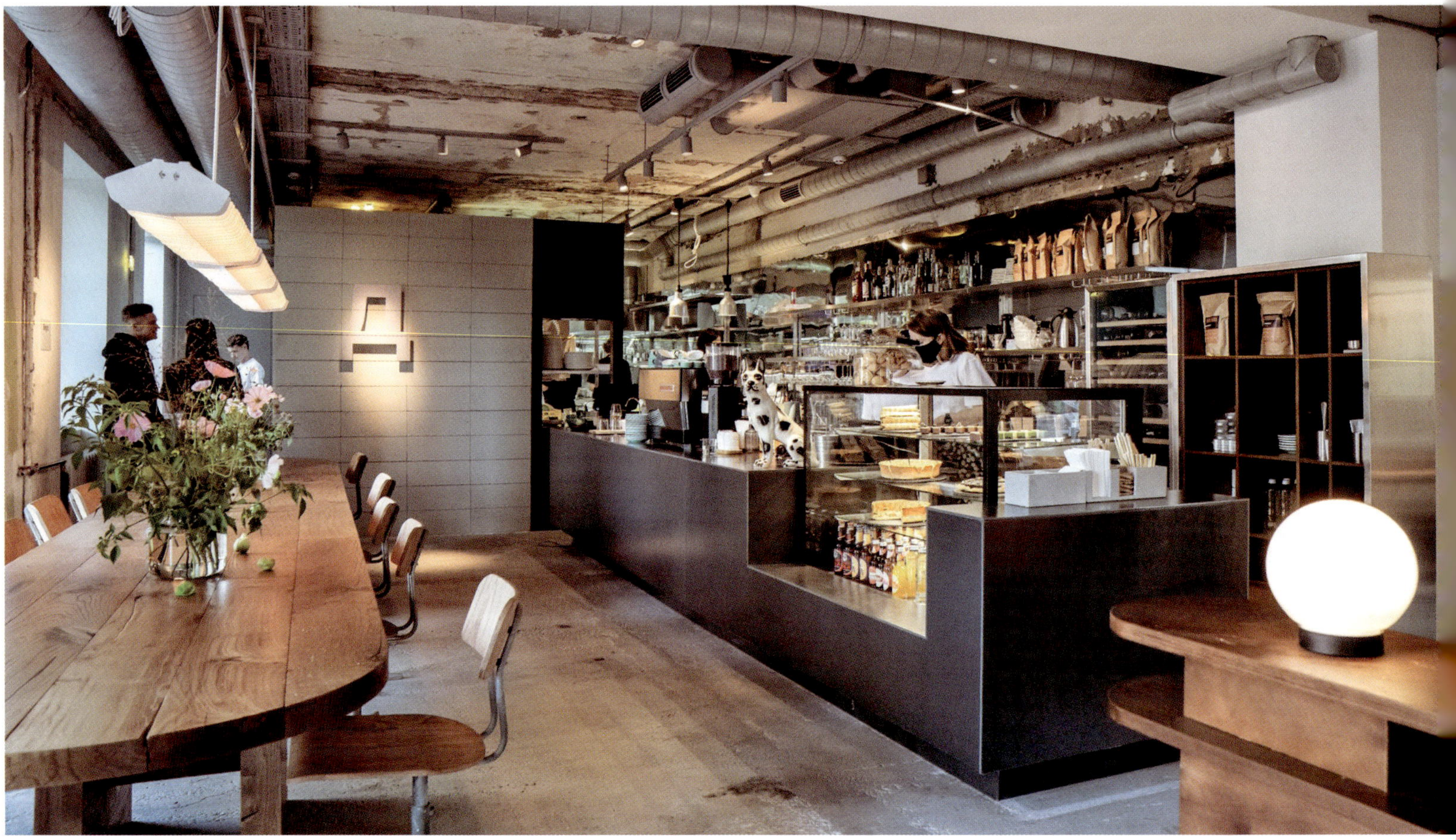

Furniture & Details

To the left of the entrance hangs a board where an image can be changed and modified daily, in a similar fashion as is done with the 'phrase of the day' board in Dyletant cafe. Ideas for illustrations are offered by the guests themselves. The six-meter communal table was created especially for this project. Swedish chairs fabricated circa 1960 were refurbished: new propro wooden seats and backrests were welded onto the original metal frame.

Through the glass ceiling of one of the bathrooms, you can see a neon sign that says "Let the whole world wait." The custom concrete sink is complemented by a swan-shaped faucet. The mural is designed and painted by artist graph0man. It features Rodin's The Thinker and a gymnast from Picasso's painting who, as the author says, are each other's doubles. One can also spot synchronized swimmers diving into a martini glass.

Lighting

Most of the cafe's lighting is achieved via spotlights. LED lamps were placed in perforated trays: in the evening the light is diffused, emphasizing the imperfect beauty of the ceiling. Above the communal table hang three lamps found at a flea market fabricated circa 1940. During their restoration, they were cleaned, sanded, painted, reassembled and welded into one single structure.

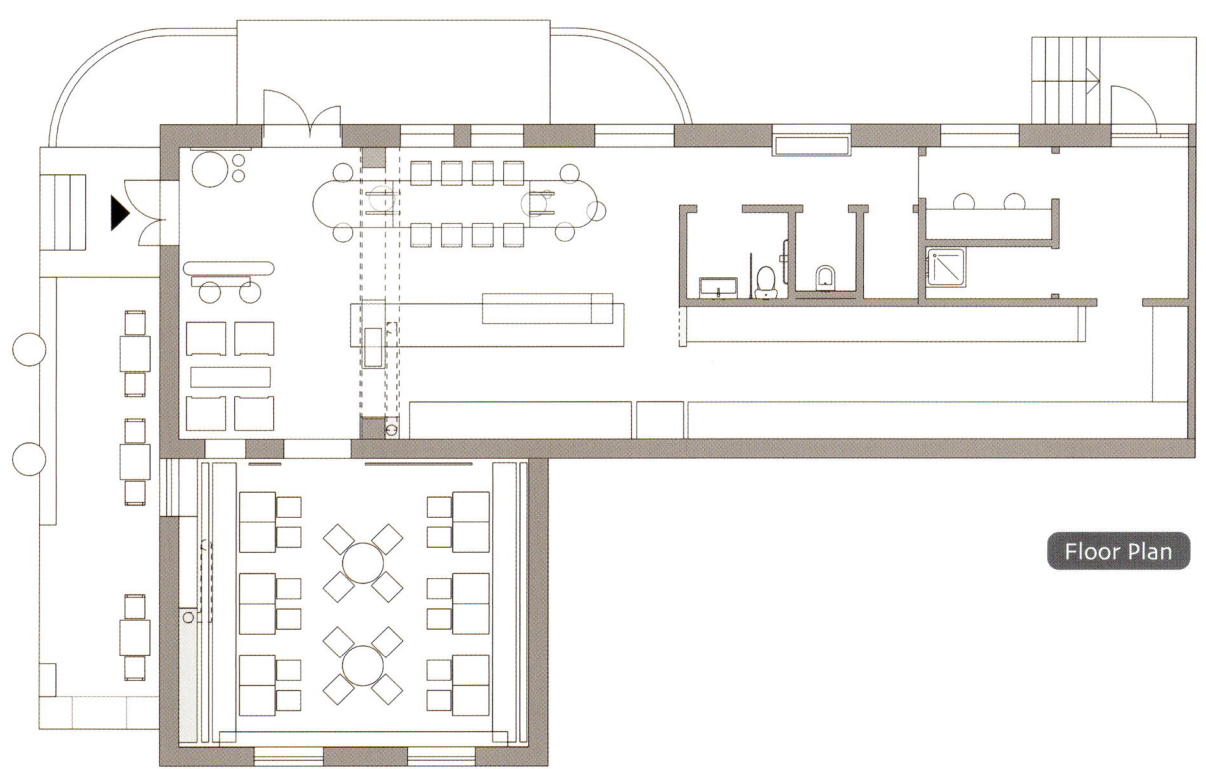

Floor Plan

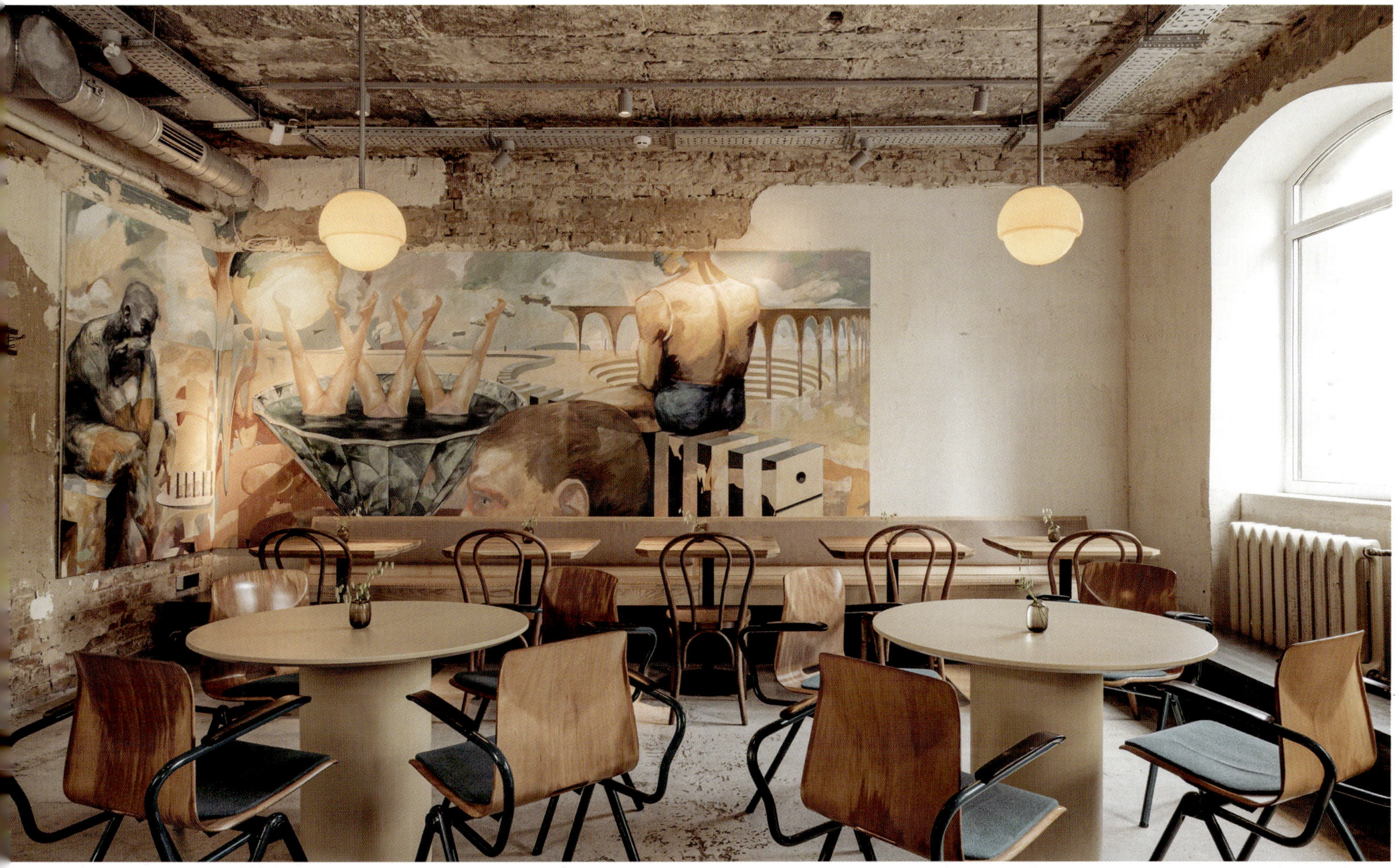

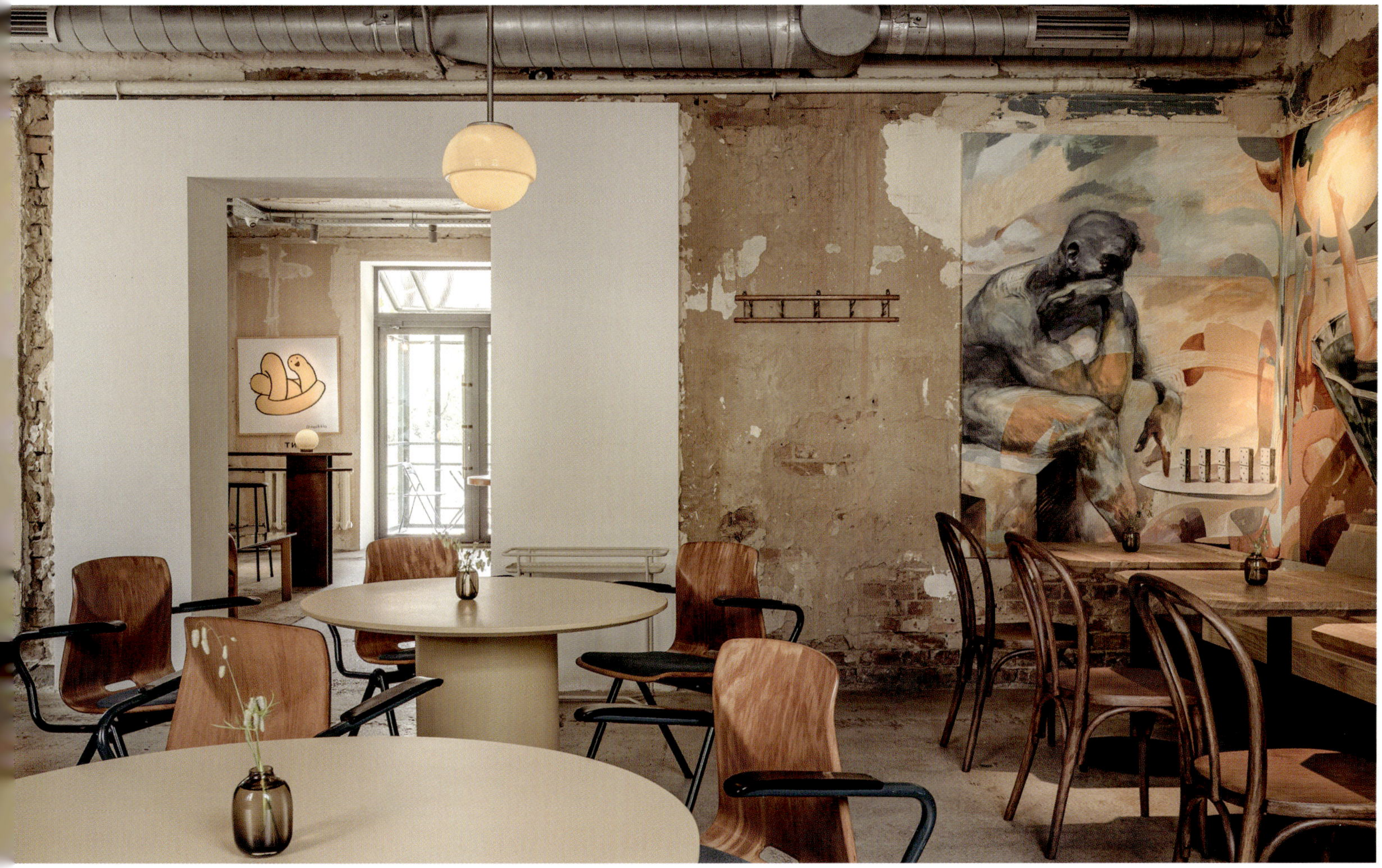

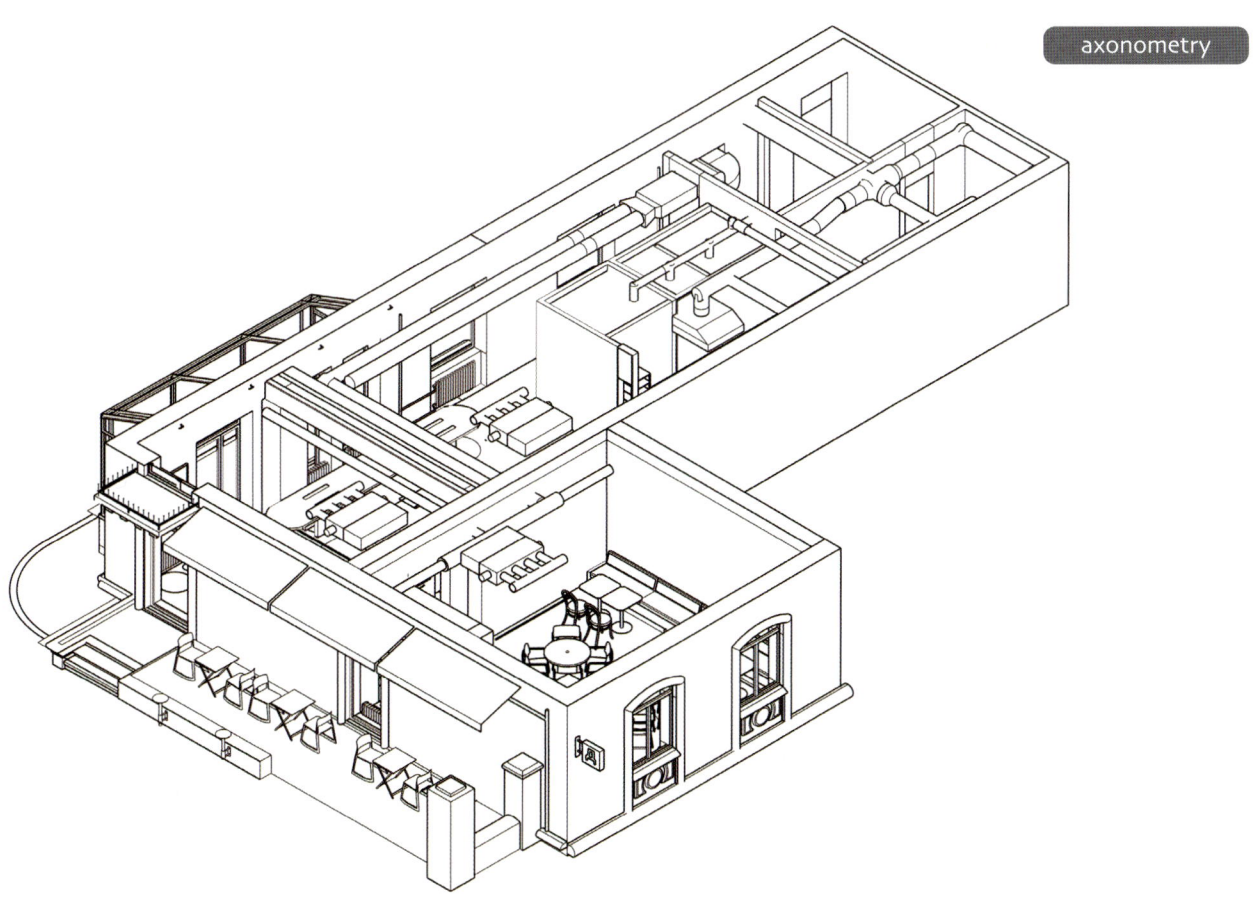

axonometry

Exterior We aimed to keep the original stucco window arches and emphasize them as much as possible. Some of the old windows were replaced with new plastic ones. To construct the entrance, we had to partially dismantle the relief on the facade of the building. We decided to recreate it on the terrace and poured identical elements out of concrete. Several seats are provided on the concrete parapets. Metal tables placed outside are the same as the ones designed for Dyletant cafe.

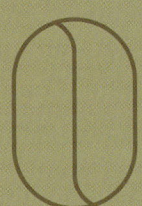 +  +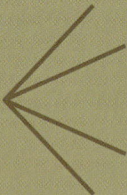

COFFEE BEAN
(SUBJECT ORIGIN)

COFFEE TUBE
(INTERIOR FOCAL POINT)

RAWR
(NAME)

RAWR Cafe
Bangkok, Thailand

RAWR

Rawr cafe?

"RAWR Cafe" is hidden in a project "in the forest" or "Naiipa Private Office Space" . This project has won world class awards. which is designed without cutting and not damaging trees This is another beautiful and peaceful project, perfect for relaxing.

Design: *OwT Branding & Design* **Homepage:** *www.behance.net/gallery/RAWR-Cafe* **Creative Director:** *Sabhat KK Rak*
Location: *Prakanong, Klongtoey, Bangkok, Thailand* **Image:** *Sabhat KK Rak*

You don't have to go to other provinces to experience nature closely with a secret cafe in the heart of Phra Khanong. that the hip-hop fans should definitely not miss! This café is hidden in the "Forest" project or "Naiipa Private Office Space". Let me tell you that this place is beautiful, shady and extremely private, like not being in Bangkok at all. The beginning of the cafe came from the owner's passion for going to cafes. So the idea was born and it was developed to become "RAWR Cafe" and the meaning of the name of the shop that many people may wonder is derived from the slang words of the roar of animals.

"RAWR Cafe" is hidden in a project "in the forest" or "Naiipa Private Office Space" . This project has won world class awards. which is designed without cutting and not damaging trees This is another beautiful and peaceful project, perfect for relaxing. Juicy with "Whiter RAWR" Signature Drink for milk coffee lovers only. In this glass, there will be softness and tenderness in every taste. The taste is lightly sweet, not greasy, cuts the throat. It also gets the taste and aroma of coffee clearly and full of words.

Design & Strategy

RAWR is about the contrast between the elegance of coffee and the power of caffeine. The main mission of RAWR is to portray the nature of drinking coffee into one simple logo.

The team started out by clarifying the basics of what the cafe is. RAWR is a cafe that specializes in coffee, located at Naiipa (meaning in the forest) Private Office Space, Bangkok, Thailand. The cafe itself is surrounded by trees which have stood resolutely for over thirty years. Therefore it is a place where people can escape from the high-rises, pollution, chaos of city life and enjoy a nice cup of coffee. The name RAWR signifies where coffee and forest intersects.

The beauty of coffee is its character. Visually; when you see a cup or someone drinking coffee, you get the sense of something simple, smart and elegant. On the other hand caffeine gives you power and energy. To visualize something that is so simple and stable yet powerful and energetic at the same time is the challenge of this project. The grid of the logo comes from the shape that symbolizes the three key elements of the cafe - the coffee bean, coffee tube and the name "RAWR" itself.

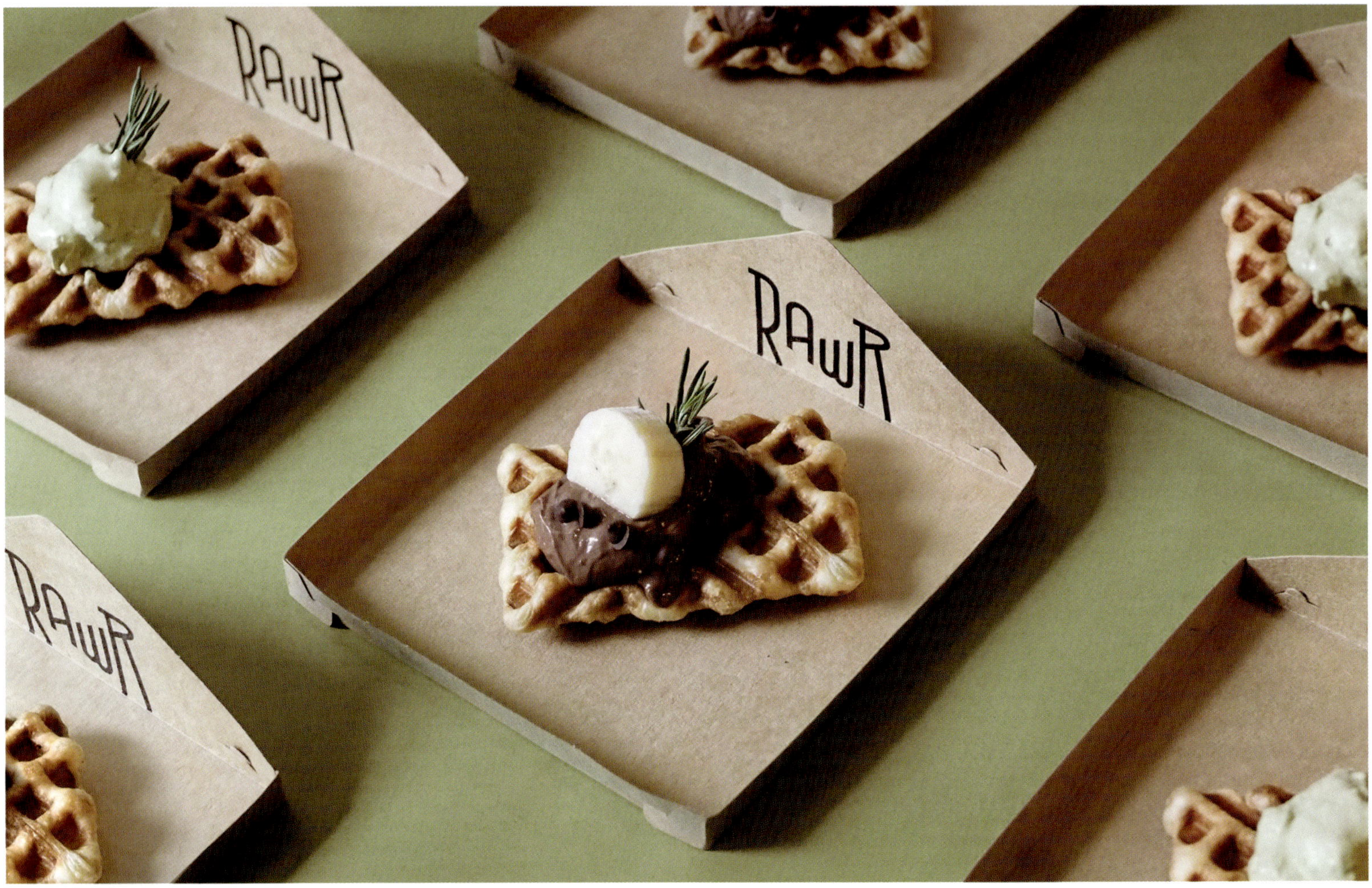

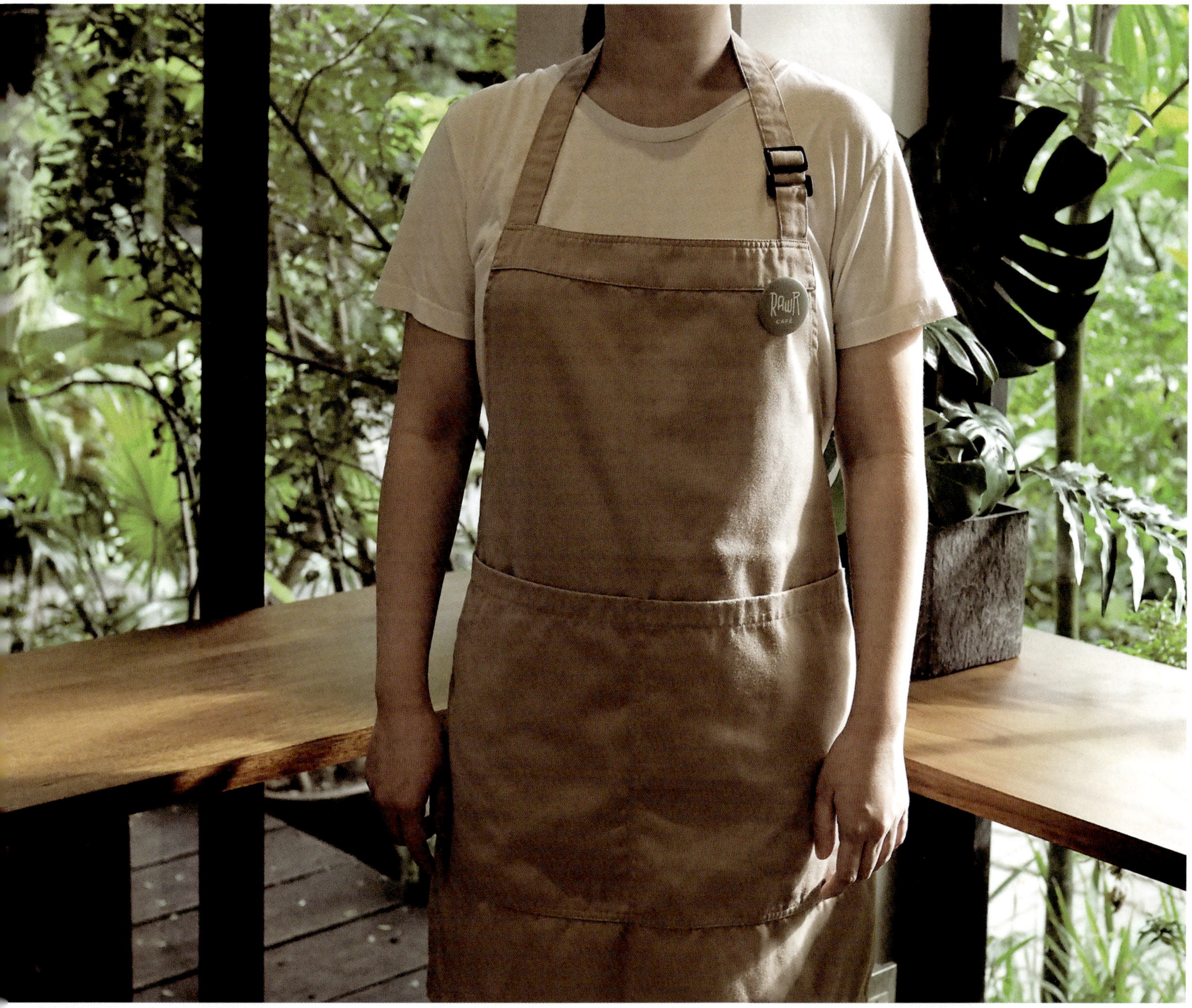
RAWR
CAFÉ

RAWR
TALK LESS
RAWR MORE
TALK LESS
RAWR MORE
BOLD TO BE

RAWR
TALK LESS
RAWR MORE

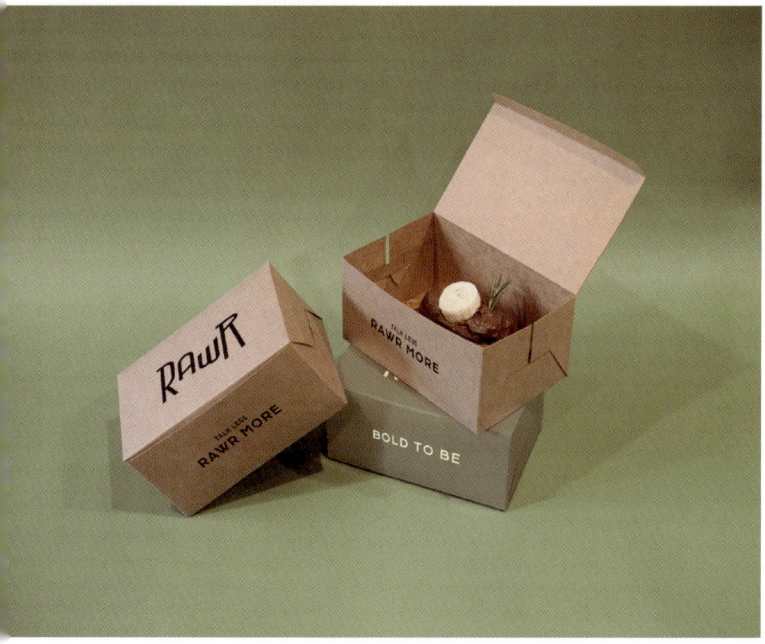

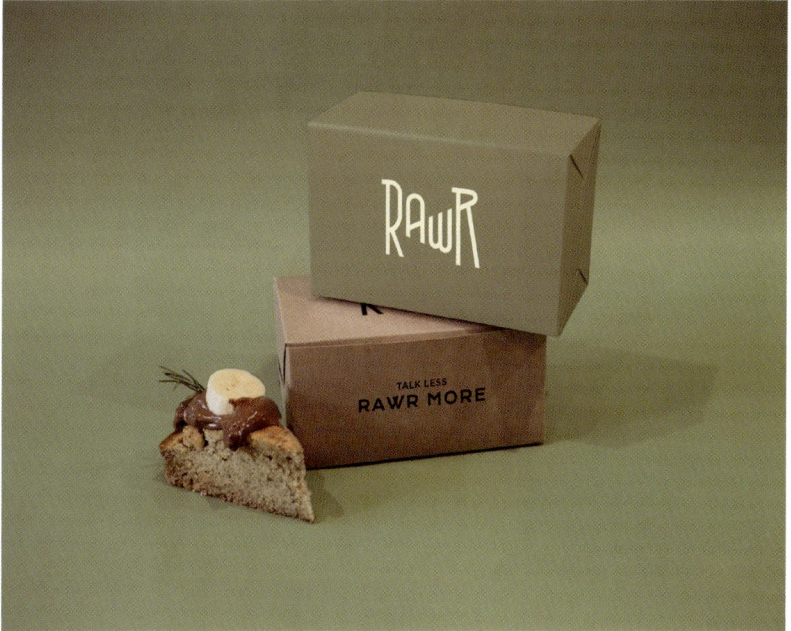

039

 EVERY CUP IS A JOURNEY

Coperaco Coffee
New York, United States

COPERACO

Concrete?

Since 1997 concrete has developed concepts in architecture, interior design, urban development and brand development. We work with a team of 50 multidisciplinary creatives for corporations and institutions. Concrete is known for branding in hospitality, retail, culture, and now, residential design. Our projects include all citizenM hotels around the world, W Hotels in London and Switzerland, Zoku Hotel in Amsterdam and the stores for Rituals and Bose, among other retailers. Currently, we are working on the masterplan for Manifattura Tabacchi in Florence. Concrete loves provoking, confusing, philosophizing, scale models, haute cuisine, burgers, and (most of all) shattering dogmas. Concrete provides solutions. No grand theories or abstract ideas — just things that work. Concrete likes to let the work do the talking.

Design: Concrete **Homepage:** www.concreteamsterdam.nl **Executive Architect, Executive Landscape Architect:** Minno and Wasko **Location:** New York, United States **Photography:** Ewout Huibers **Creative Direction:** Alexandros Gavrilakis

---

Coperaco is a boutique coffee roaster company operating in New York, Mexico and Colombia. Their new cafe in New York creates an idyllic sanctuary where patrons can savor the high-quality, carefully-roasted brew. The coffee created by Coperaco—which can be found at some of the world's best restaurants and hotels—is hand-roasted in small batches, and focused on the farming and harvesting of the fruit, creating a connection to the earth.

AG was asked to help create their brand elements, starting from their most extrovert point. Their coffee bags. Our role was to re-design their 12 & 35 oz coffee bags, giving an elegant new life and a desirable character to the product, bringing to life the components that holistically make up the Coperaco brand. The illustration wrapping the bags take us a magical journey discovering the world's best coffee plantations, local farmers and beautiful lands, rewarding excellence & craftsmanship. The coffee tree is there to remind us that producers are the heart of what makes coffee great.

Coperaco, a celebrated New York-based roaster, opens its first U.S. cafe in Harrison, New Jersey—and it's a stunner. Serving as the ground-floor lobby of Harrison Urby, a lifestyle-oriented rental complex, Coperaco's first U.S. cafe is dripping with lush foliage. Founder Johan Pesenti collaborated closely with Amsterdam-based architecture and interiors firm Concrete to create an idyllic sanctuary where patrons can savor the high-quality, carefully-roasted brew.

"We wanted to have an open space, lit by natural light, where you have the feeling of being in a greenhouse," Pesenti explains. The coffee created by Coperaco—which can be found at some of the world's best restaurants and hotels—is hand-roasted in small batches, and focused on the farming and harvesting of the fruit, so creating a connection to the earth was essential to the design of the cafe. A vertical batten dripping with fronds encloses the heavy, under-lit marble coffee bar. Mosaic floor tiling grounds the space as barn framework fuses the indoor and outdoor elements. Airy, cascading lighting melds with hanging ivy throughout the space. Adding further to the cafe's outdoor theme is a two-story tree house built inside the space. Made of wood with a wall of braided rope, the structure adds cozy seating, a library, and a fireplace. Furnished with sleek midcentury-style furniture, it's a transformative space. One certainly does feel as if they are in the Garden State.

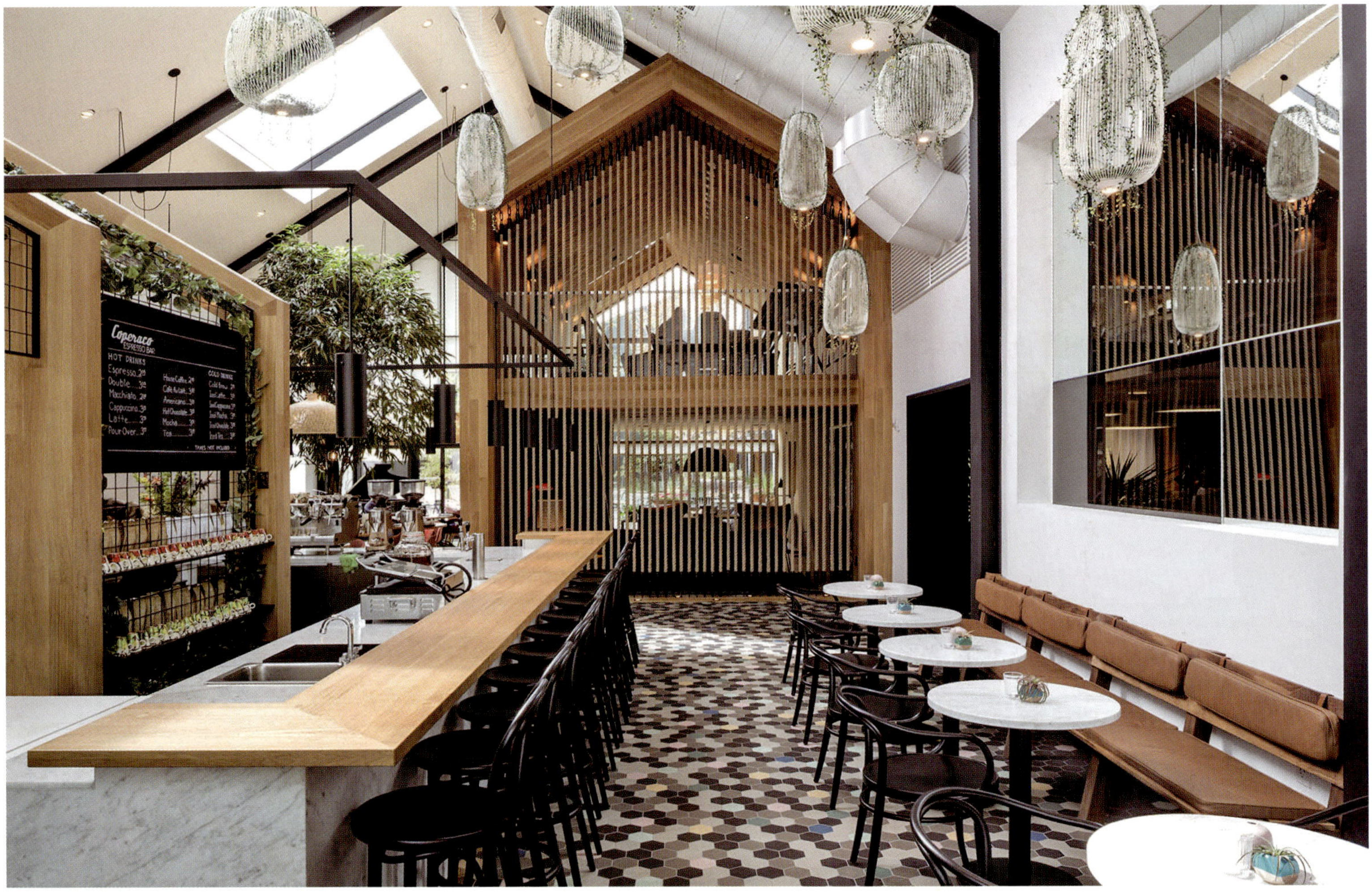

Myfresh Café

Panchkula, India

Come for a refreshing experience at
The Biggest Gourmet Store of PANCHKULA

 **Organically Grown** 100+ Variety of **Fruits & Veggies**

Responsibly Sourced 5000+ Range of **Imported & Indian Groceries**

 **Healthy & Concious** 100+ Delights at **Kitchen & Café**

myfresh

Loop Design Studio?

We are a B Corp certified boutique architecture and design studio based in Costa Rica, committed to creating spaces that are purposeful, personal, and built in the best way possible. We put humans and nature at the heart of our design process, looking to build spaces that connect people, the natural world, and the built environment. Our focus is on project development that has a positive impact on quality of life as well as on a project's ecosystem.

Design: Loop Design Studio **Homepage:** www.loopdesign.studio **Team:** Suvrita Bhardwaj, Nikhil Pratap Singh, Sargam Sethi, Simran Chawla....
Location: Panchkula, India **Photography:** Purnesh Dev Nikhanj **Area:** 600 m²

"The curve is the line of the Gods." Myfresh Café is an attempt to create a bold and sinuous space that mimics natural form unequivocally and unabashedly. It is an experimental endeavor that explores the minimal aspect of seamless forms.

Architect Charles Deaton said, 'If people do not have angles, then we should not live in boxes. The perishable grocery super-brand called for an unconventional, fluid, and adhesive design for its restaurant section. The idea was to create a homogenous volume characterized by curvilinear ceilings and statuesque Dholpur stone slivers over the blank canvas. The free-form oak furniture adds a second meandering layer that enhances the zaftig volume.

The third accent layer of green and floor lamps gives the space a vertical impetus and depth. The fierce and striking dynamism of the space is consciously muted through a monotonal palette of beige and fawn shades. MyFresh is a contemporary outlet that reflects its patrons' vibe and tries to conjure magic through its buxom imagery.

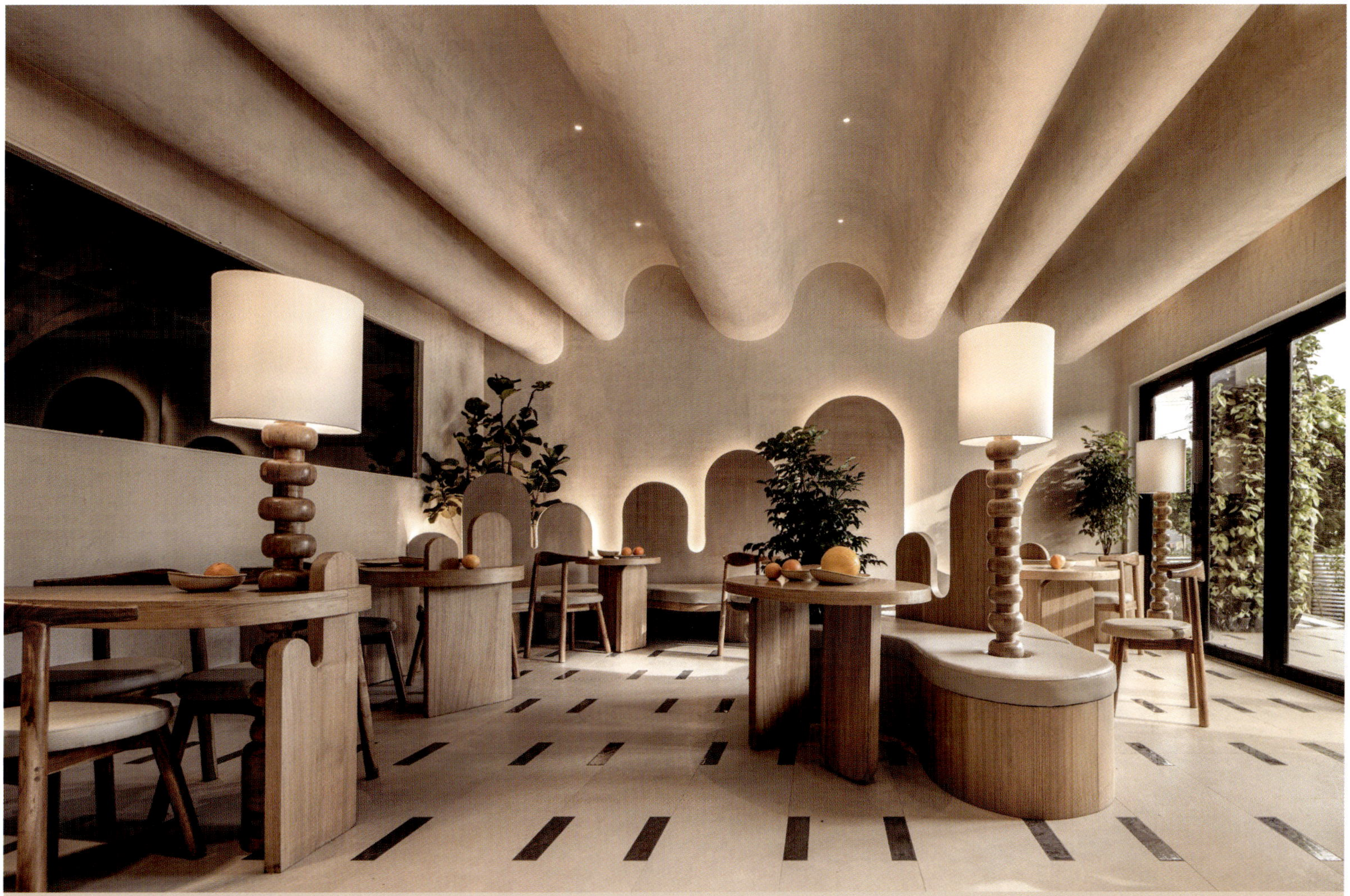

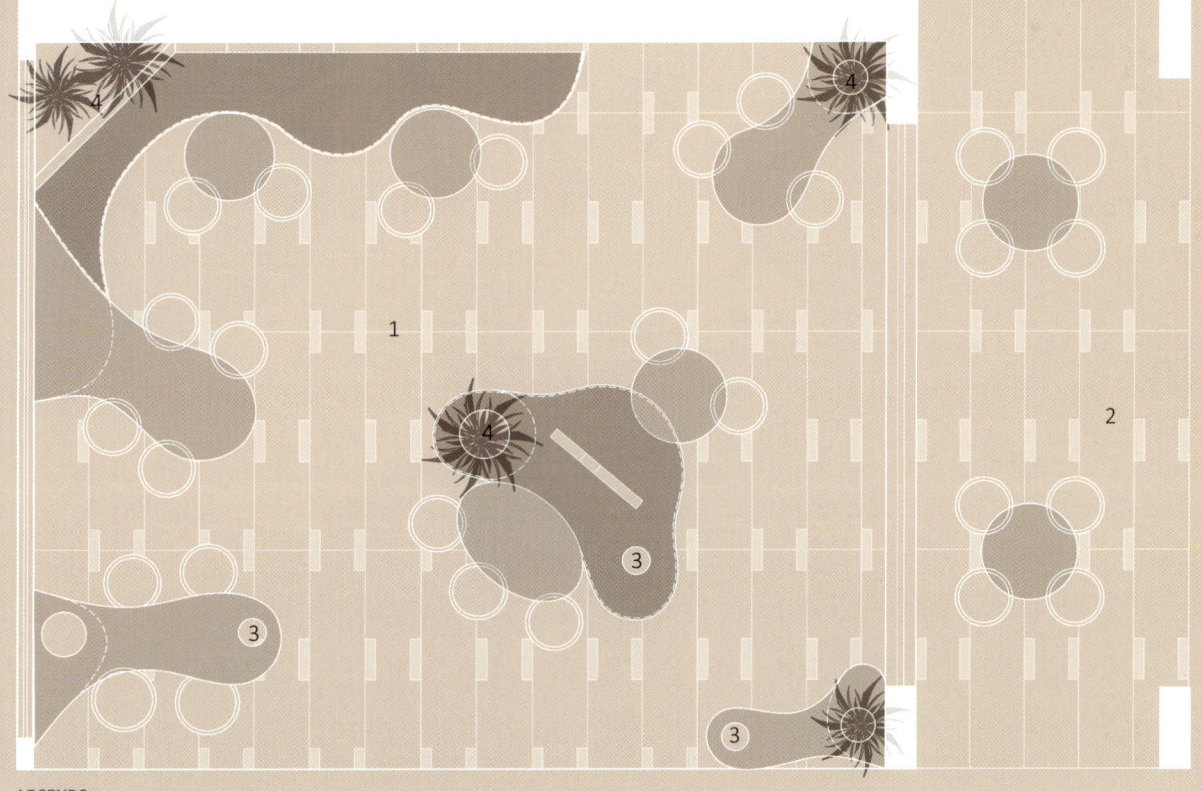

LEGENDS :
1. Indoor Seating
2. Outdoor Seating
3. Cutout for Lamp Posts
4. Planters

FLOOR PLAN
MYFRESH CAFE

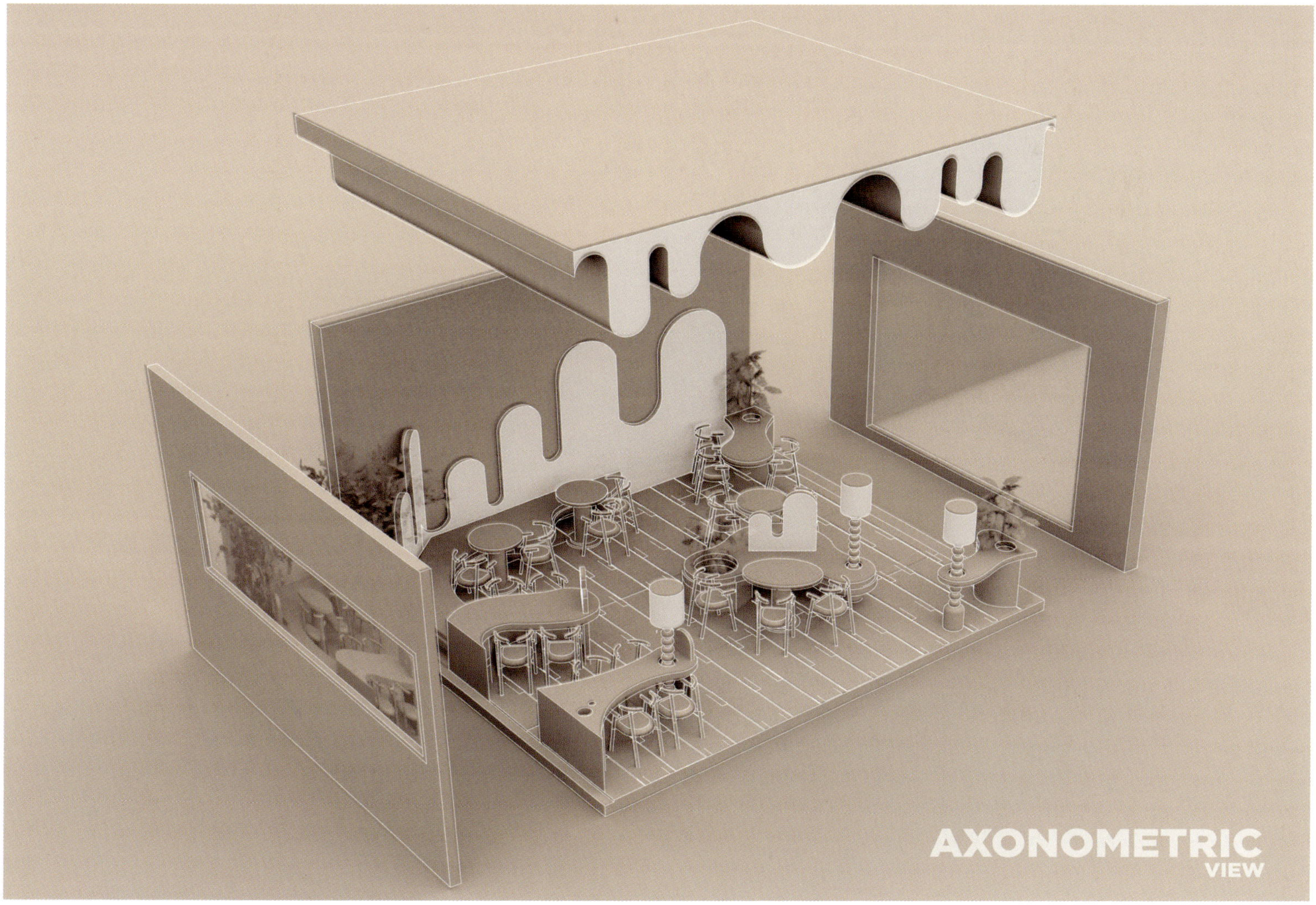

AXONOMETRIC VIEW

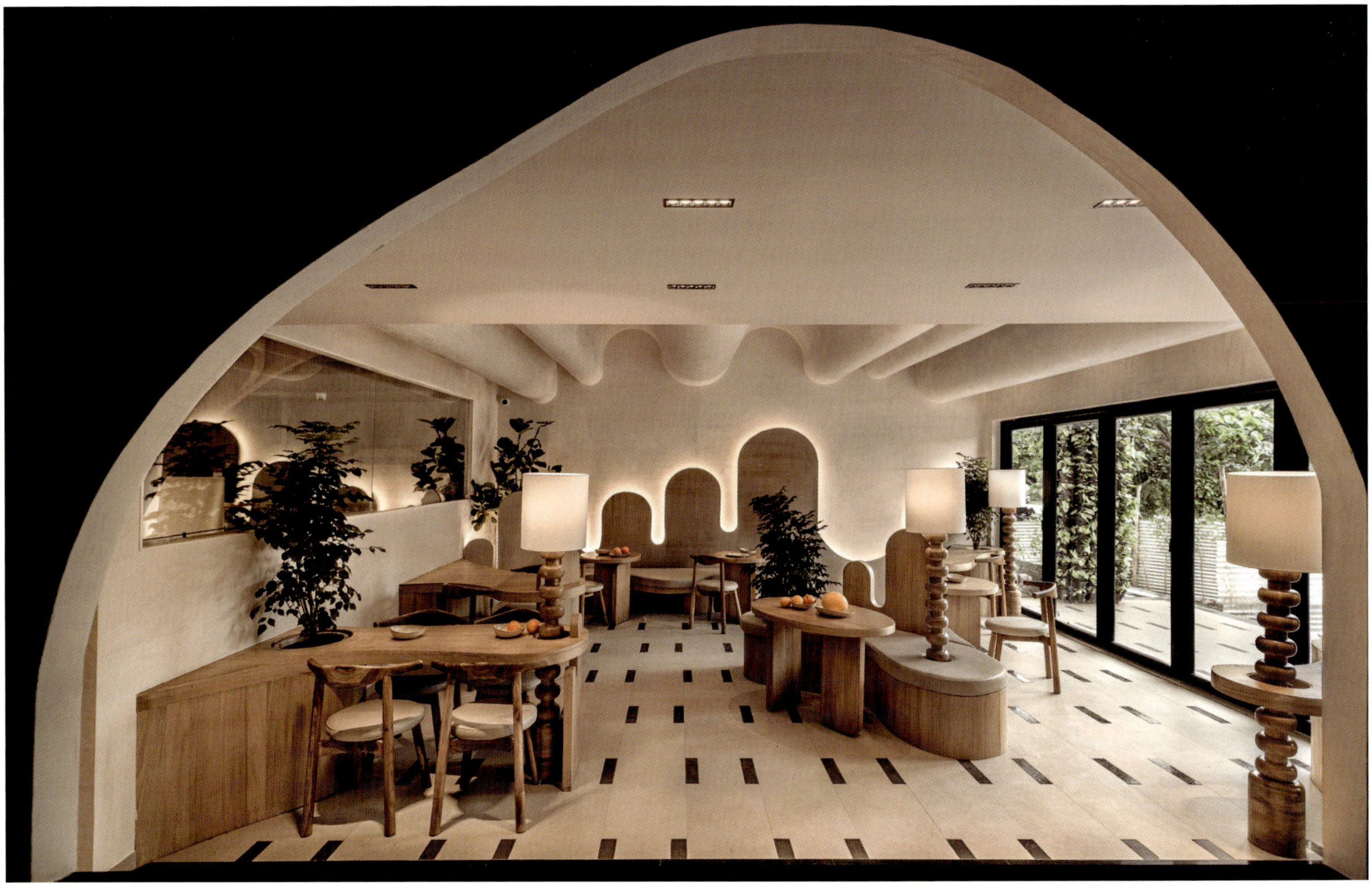

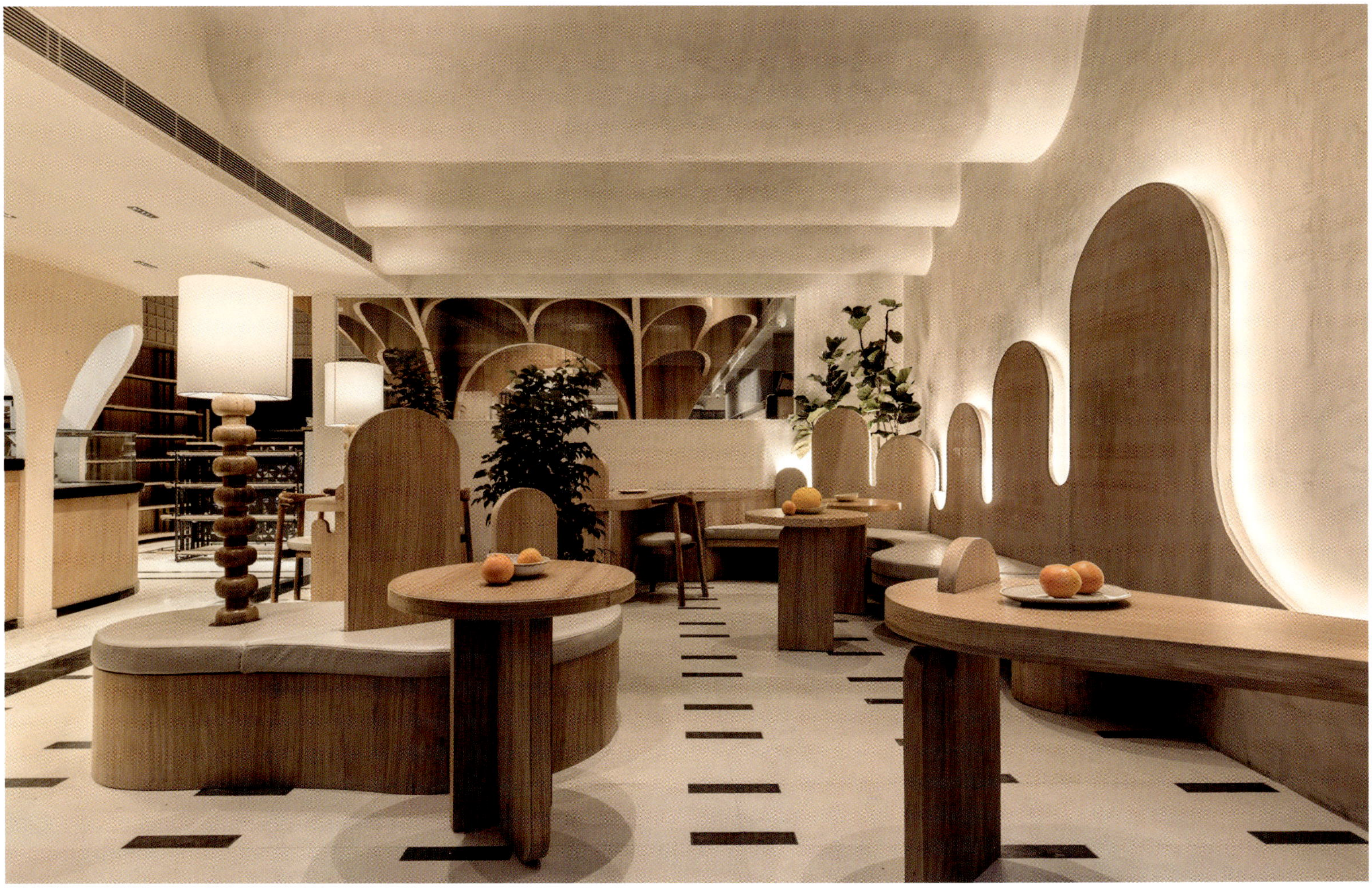

Blue Bottle Coffee

Kyoto Kiyamachi Cafe Kyoto, Japan

BLUE BOTTLE COFFEE

GIANT STEPS

Notes of cocoa, toasted marshmallow, graham cracker

WHOLE BEAN BLEND

NET WT 6 oz (170 g)

BLUE BOTTLE
COFFEE

Schemata Architects?

Jo Nagasaka established Schemata Architects right after graduating from Tokyo University of the Arts in 1998. He established the shared creative office "HAPPA" in 2007. Currently he has an office in Kitasando, Tokyo. Jo has extensive experience in a wide range of expertise from furniture to architecture. His design approach is always based on 1:1 scale, regardless of what size he deals with. He works extensively in Japan and around the world, while expanding his design activity in various fields.

Design: Jo Nagasaka, Schemata Architects   **Homepage:** http://schemata.jp   **Branding:** Pearlfisher
Location: Kyoto, Japan   **Photography:** Takumi Ota   **Area:** 123 m²

Former Rissei Elementary School, built along Takase River in Kiyamachi, Kyoto in 1927, was fully renovated and reopened as a commercial complex comprising a hotel and shops among others. The third Blue Bottle Coffee cafe in Kyoto opened in a space near the front entrance to the hotel. In terms of color, the renovated building is basically composed of white walls in common areas and black window frames. Considering Blue Bottle Coffee's tendency to use few colors, the basic color palette would be black and white but we felt it does not match the brand personality.

We came up with an idea of adding another color to the black and white color palette, namely green taken from a range of green colors in the surroundings including leaves of cherry trees along Takase River and dark green blackboards used in the classrooms, and used green throughout the interior. As a result, the room is enlivened by varying shades of green, and various pieces of furniture composed of materials in different shades of green resonate with the interior. In addition, we designed an island-type workshop counter with a grid-based convertible countertop and a built-in sink and drain system, which we proposed to Blue Bottle Coffee as a new type of furniture to explore new ways of their merchandise promotion.

Pearlfisher has created a new look for Blue Bottle Coffee's New Orleans Iced Coffee. Though many iced coffee brands already crowd the shelves, Blue Bottle's offering is unique. Now bringing their coffee expertise out of their cafes, Blue Bottle's New Orleans Iced Coffee is cold brewed for 18 hours with organic roasted chicory, sweetened with organic cane sugar and cut with organic whole milk. The challenge for the brand was how to share this unique offer and extend the intimate Blue Bottle experience from carefully curated cafes to a carton, reaching a larger audience and standing out on shelves in mass retailers.

To help Blue Bottle surmount this challenge, Pearlfisher created a new design for the brand that stays true to the heart of Blue Bottle Coffee and the Founder, James Freeman's original vision, at once challenging coffee category cues while remaining modest and utilitarian. Tess Wicksteed, Executive Vice President at Pearlfisher commented, "The challenge when working with a very simple and pure iconic brand is how to introduce a secondary visual language with depth that doesn't complicate the purity of the existing design. Structure is one way to do this. The milk carton allowed us to play up the brand's iconic equities without losing the crafted feel of the brand."

WINTER
BLOOMS
BLEND
NOTES OF
MILK CHOCOLATE
BLACK CHERRY
NUTMEG
WINTER
BLOOMS
SINGLE ORIGIN
Ethiopia Gedeb
METAD Chelbesa Natural
NOTES OF
BLUEBERRY
JASMINE
YUZU
A GIFT OF
BLUE BOTTLE
COFFEE

BLUE BOTTLE
HOLIDAY YOKAN

BLUE BOTTLE COFFEE
INSTANT
COFFEE

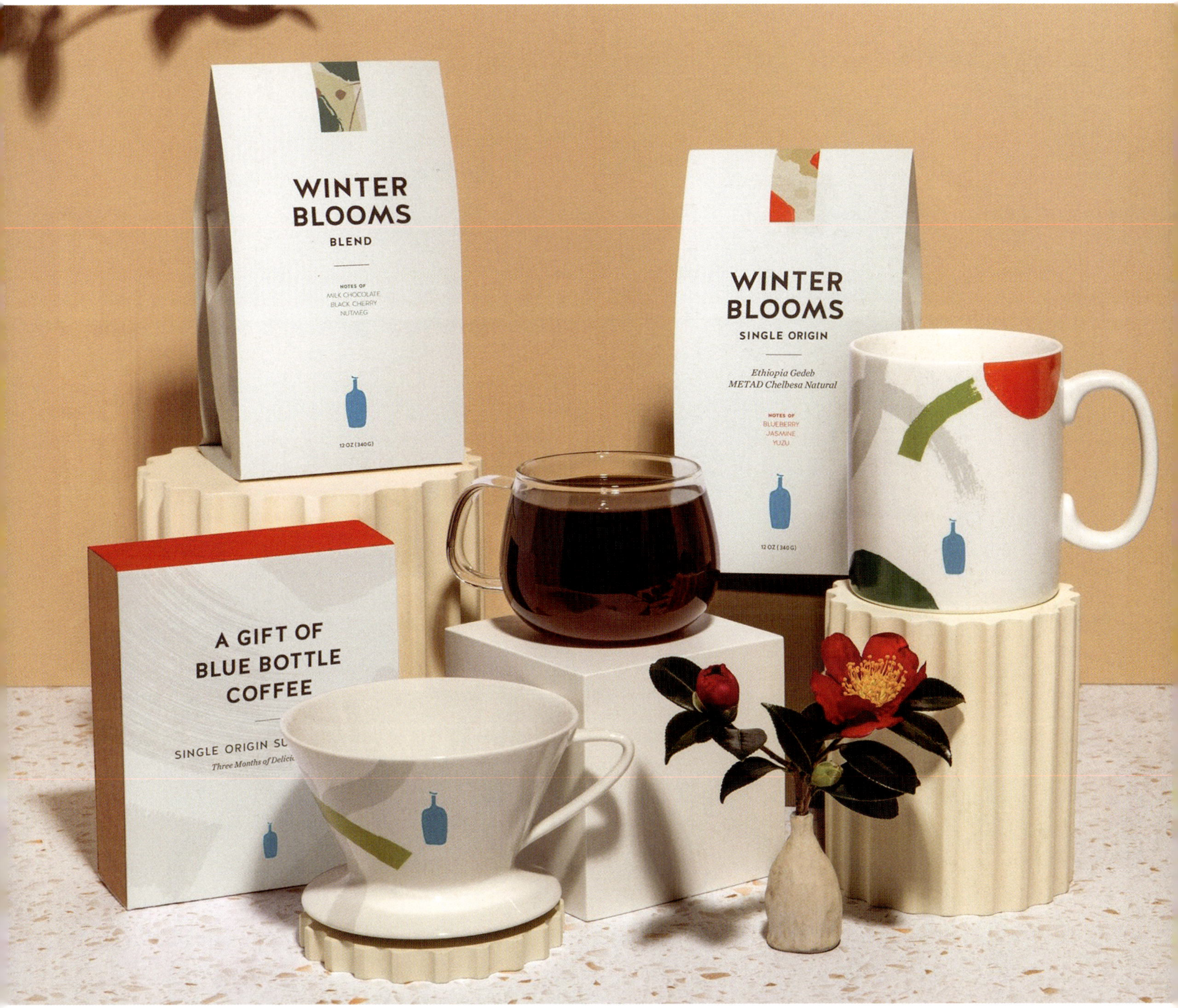

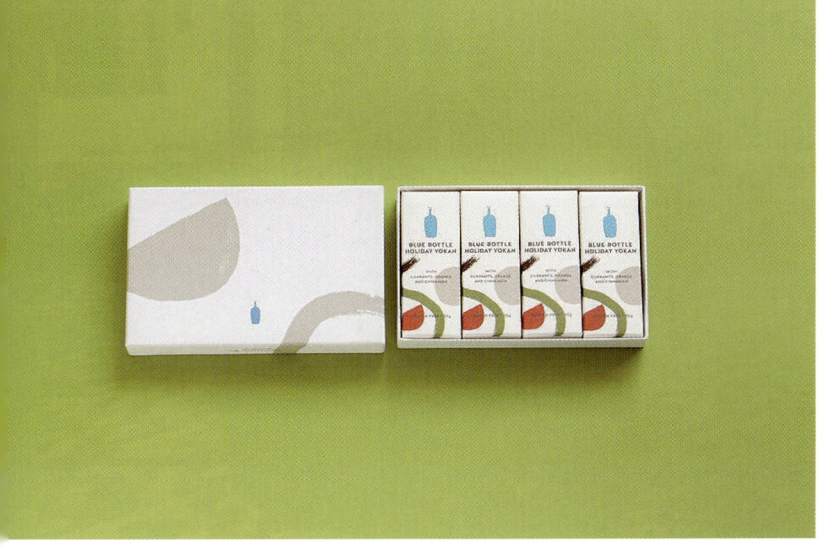

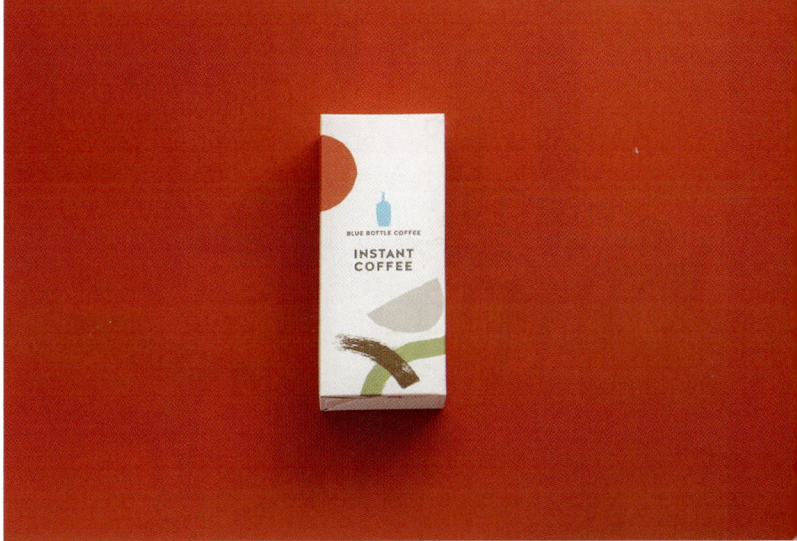

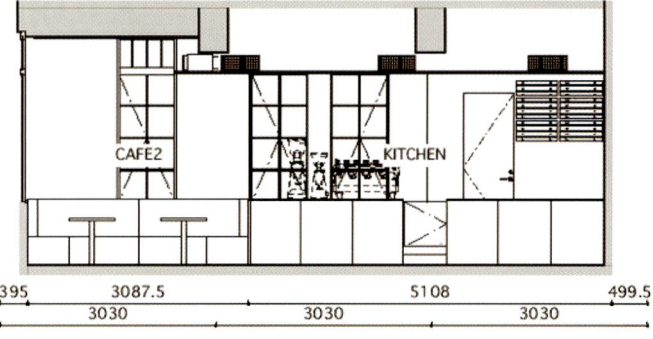

Section A 1:100

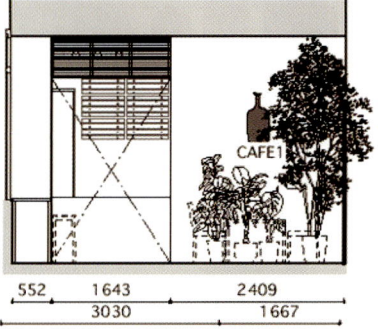

Section B 1:100

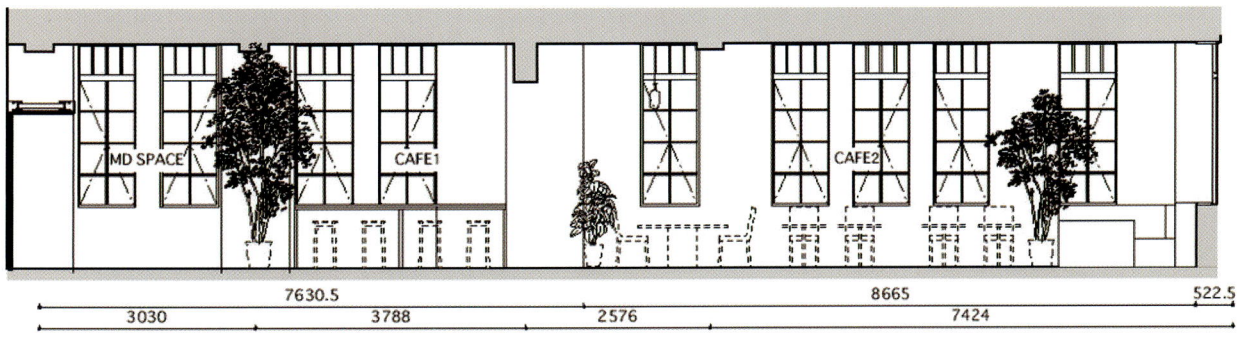

Section C 1:100

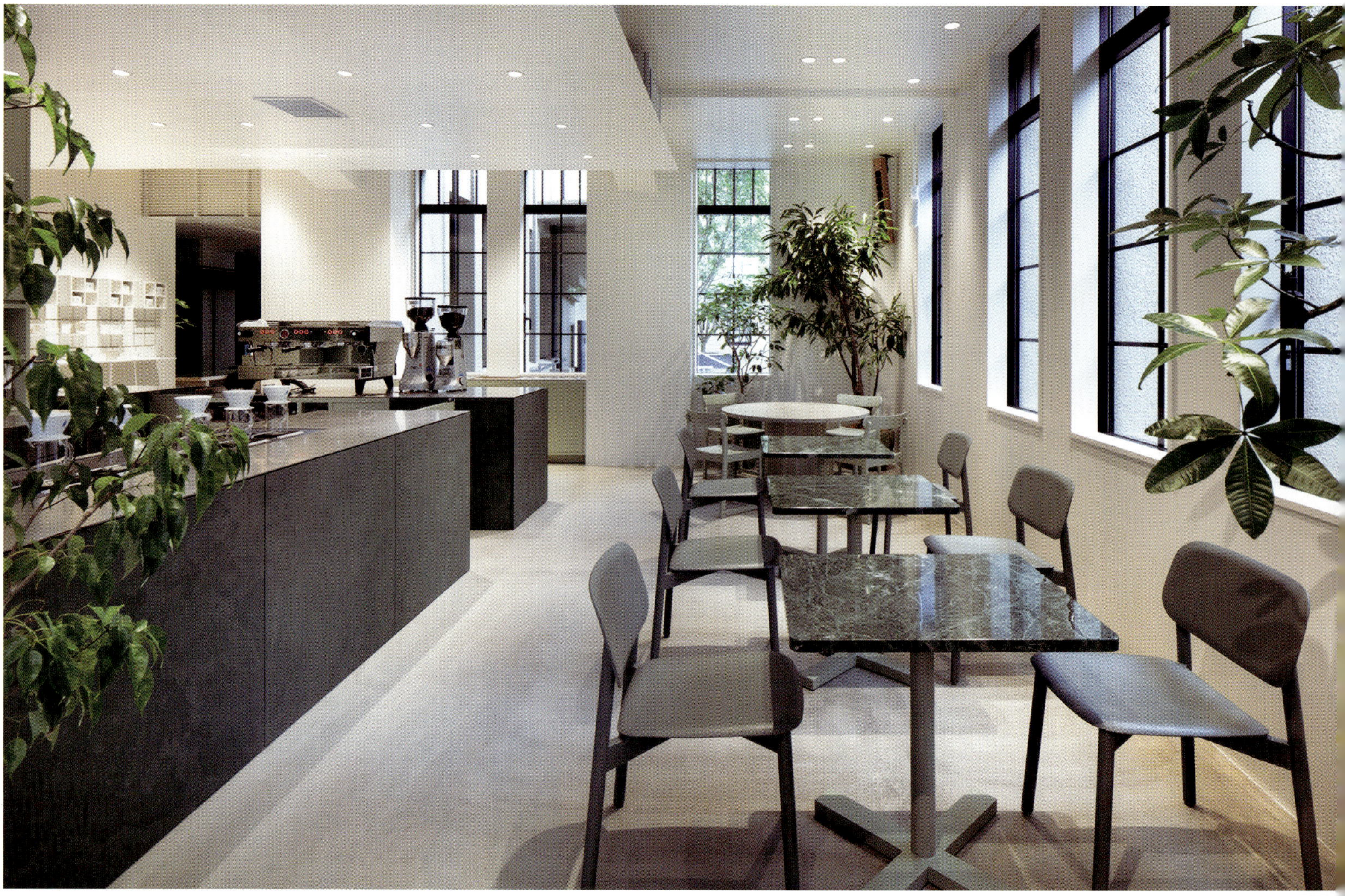

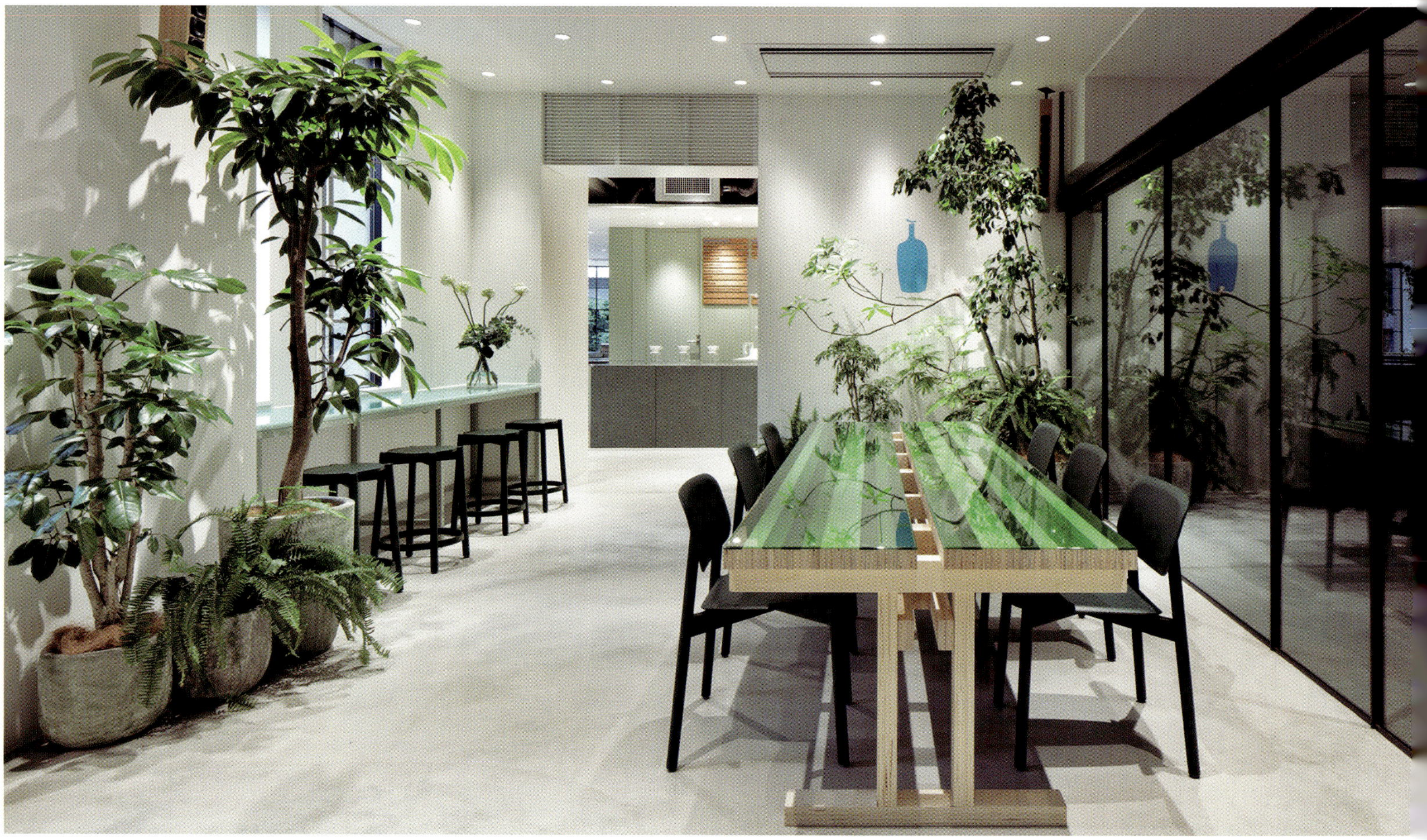

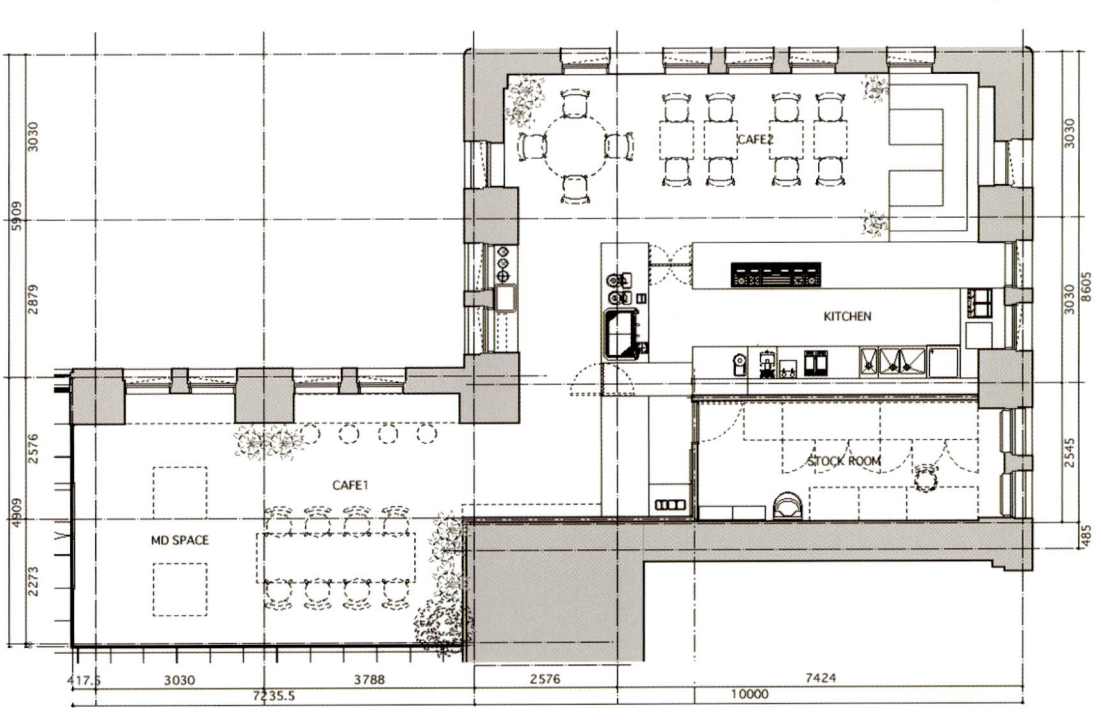

Floor Plan 1:100

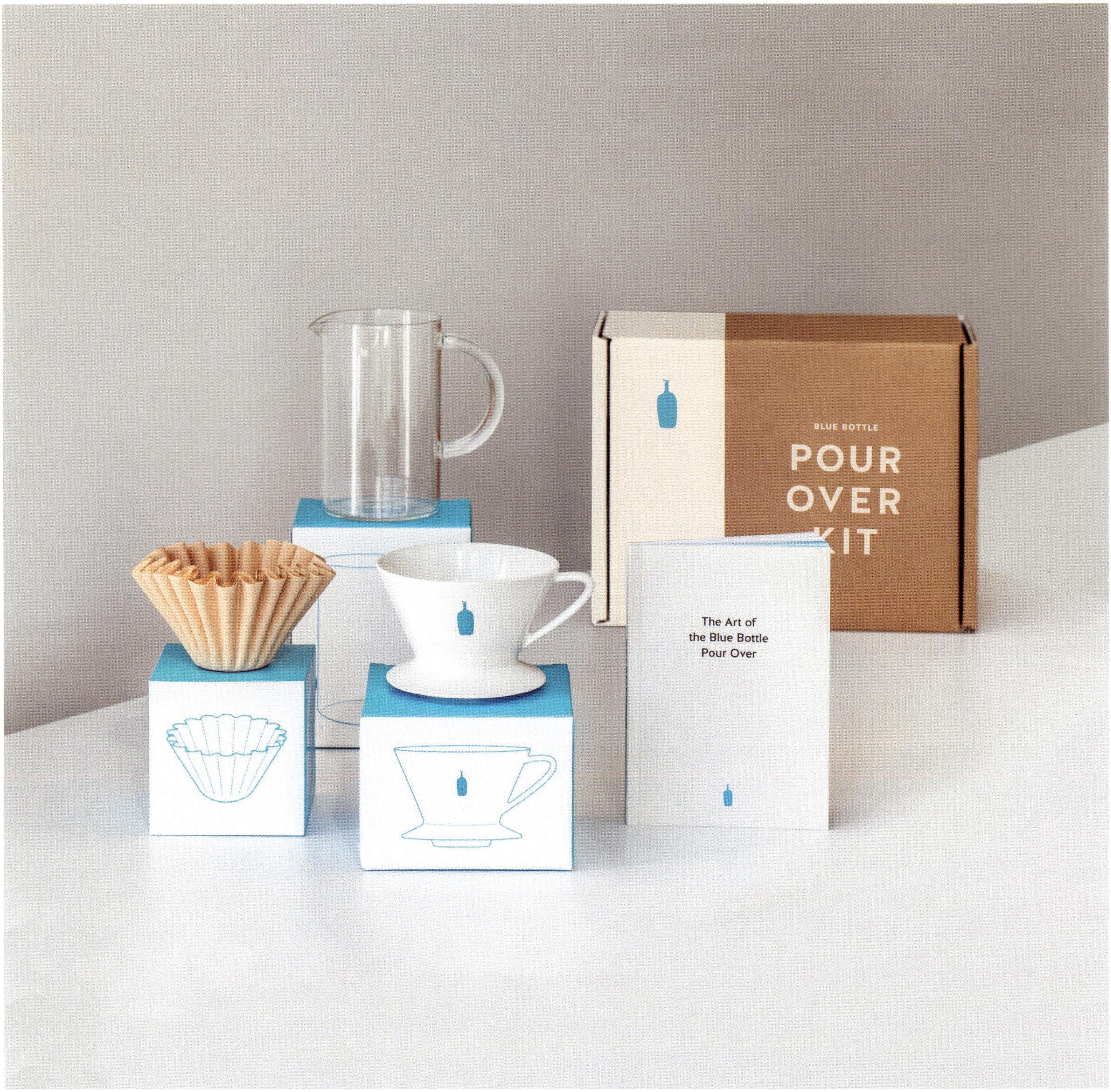
BLUE BOTTLE
POUR
OVER
KIT
The Art of
the Blue Bottle
Pour Over

BLUE BOTTLE COFFEE
GIANT STEPS

FIRST LIGHT
SINGLE ORIGIN
RENEWAL
BLEND

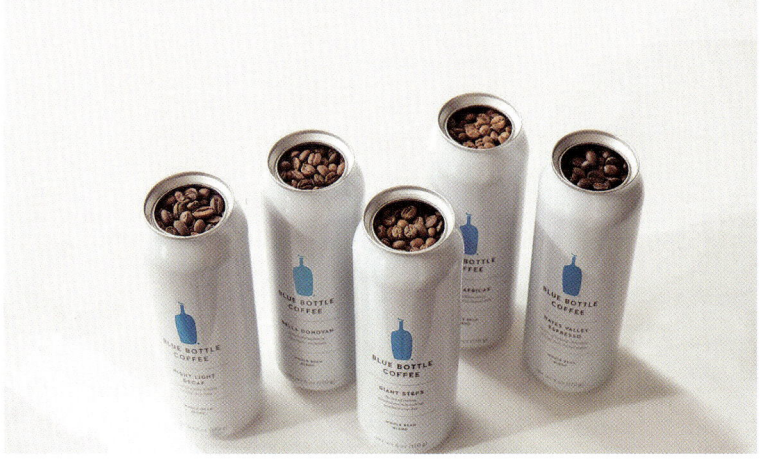

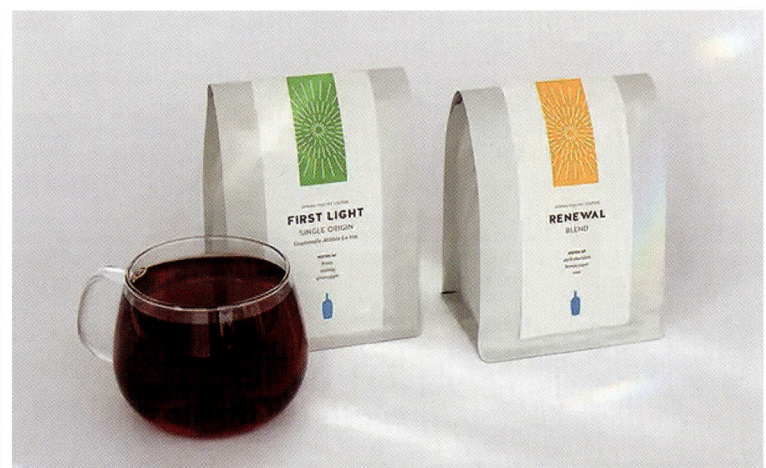

BLUE BOTTLE COFFEE
Single Origin
INSTANT

USDA ORGANIC
BLUE BOTTLE
COFFEE
COLD BREW
8 FL OZ (236ml)
PERISHABLE. KEEP REFRIGERATED.
NEVER HEATED

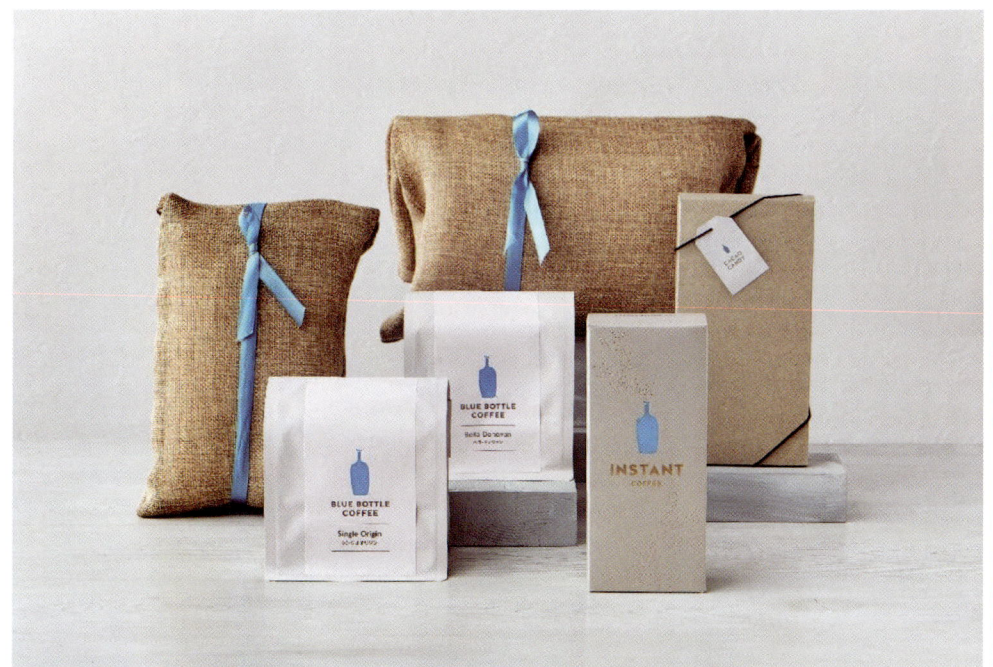

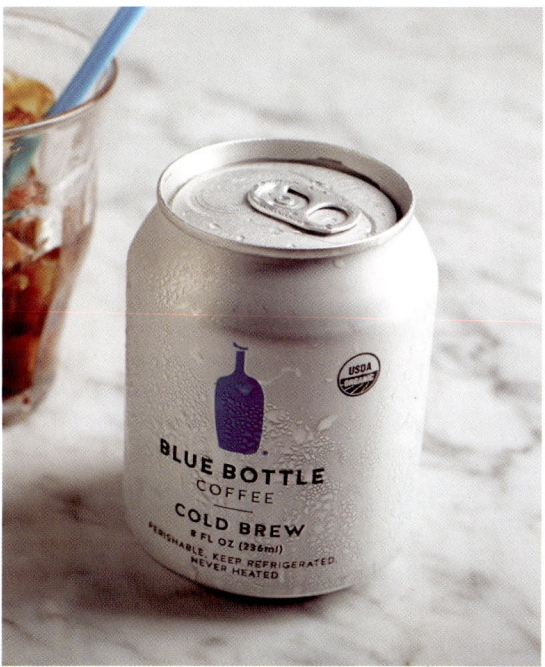

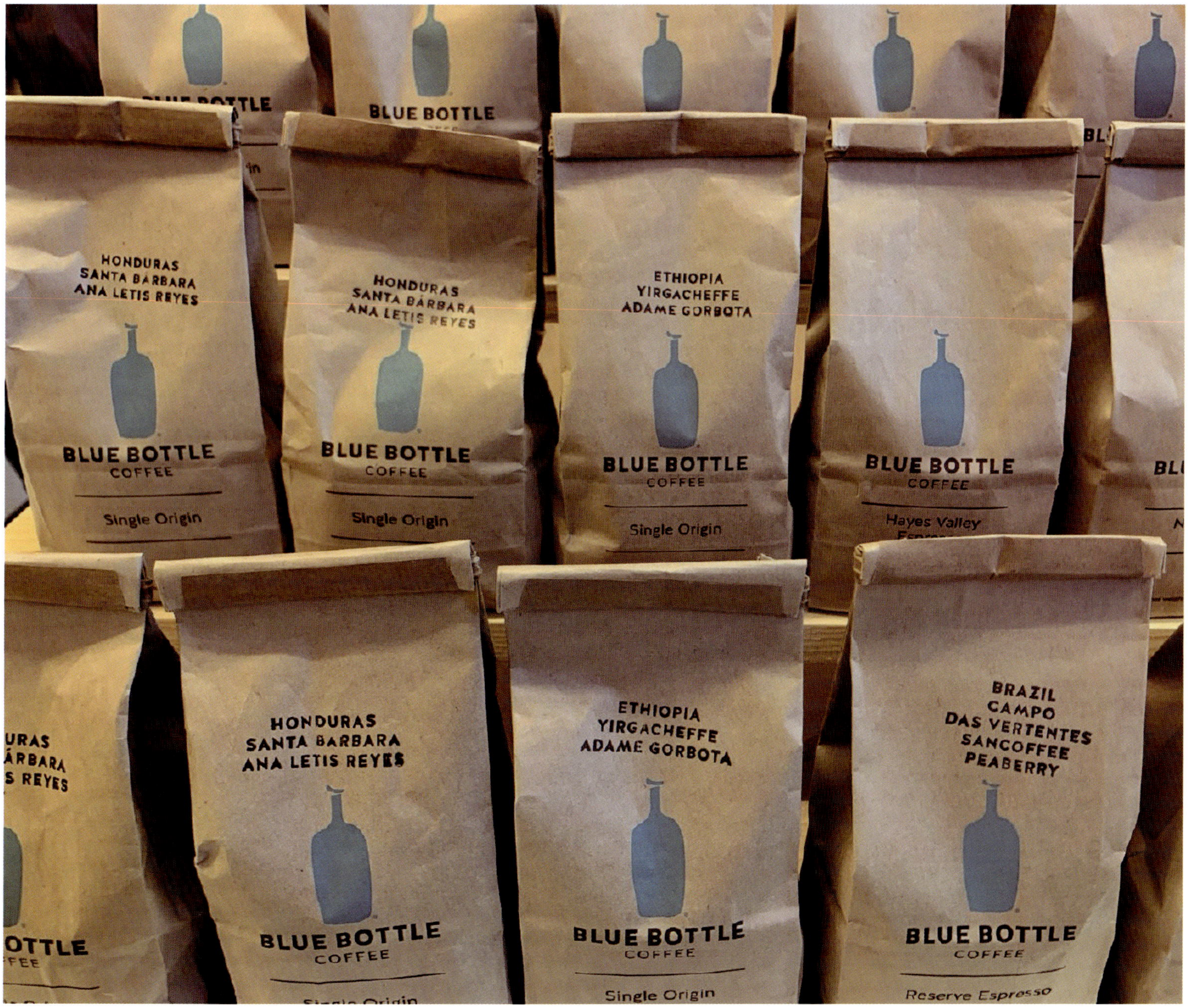
HONDURAS
SANTA BARBARA
ANA LETIS REYES
BLUE BOTTLE
COFFEE
Single Origin
ETHIOPIA
YIRGACHEFFE
ADAME GORBOTA
BLUE BOTTLE
COFFEE
Single Origin
BLUE BOTTLE
COFFEE
Hayes Valley
BRAZIL
CAMPO
DAS VERTENTES
SANCOFFEE
PEABERRY
BLUE BOTTLE
COFFEE
Reserve Espresso

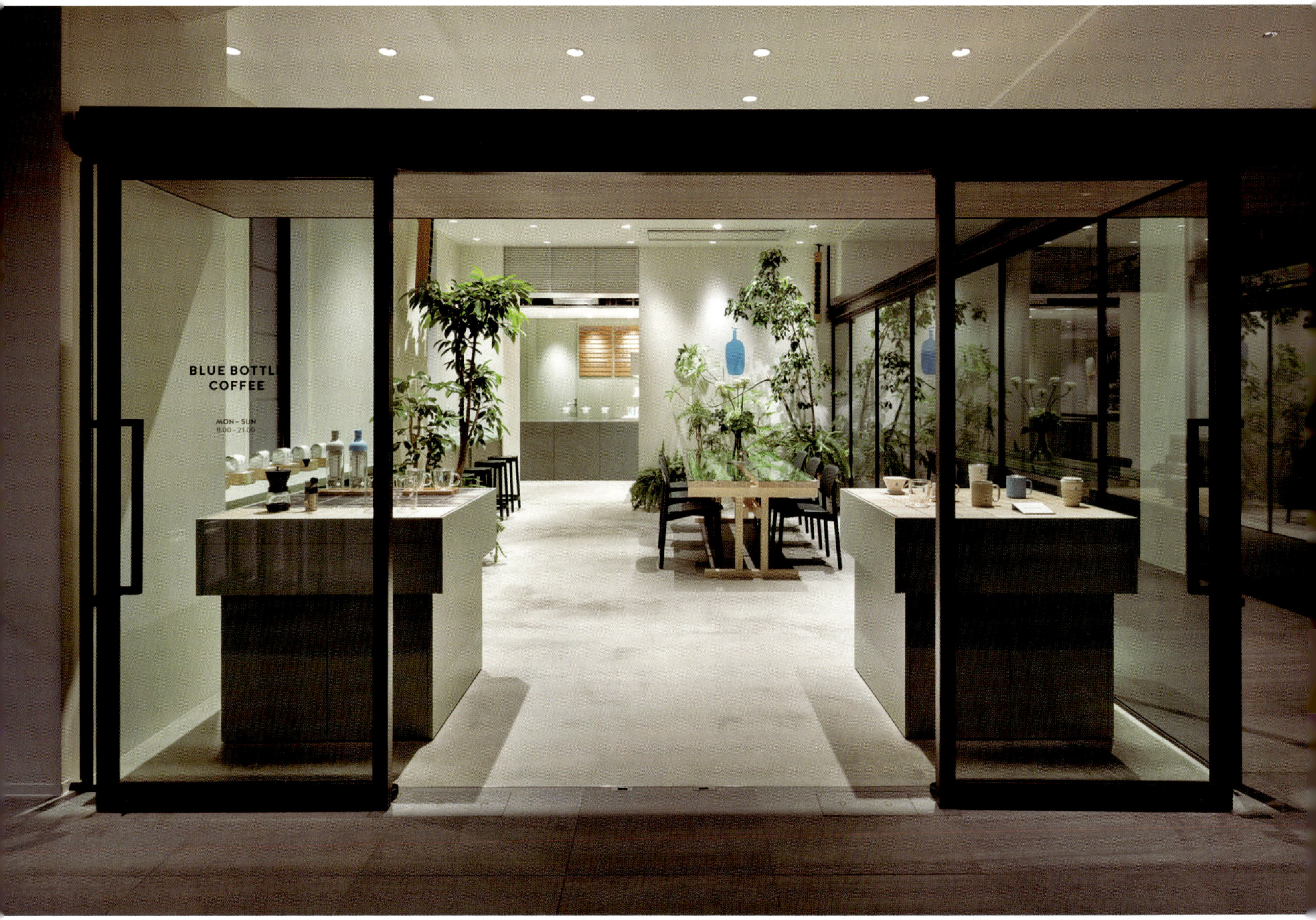
BLUE BOTTL
COFFEE
MON – SUN
8.00 - 21.00

DRINK WARE

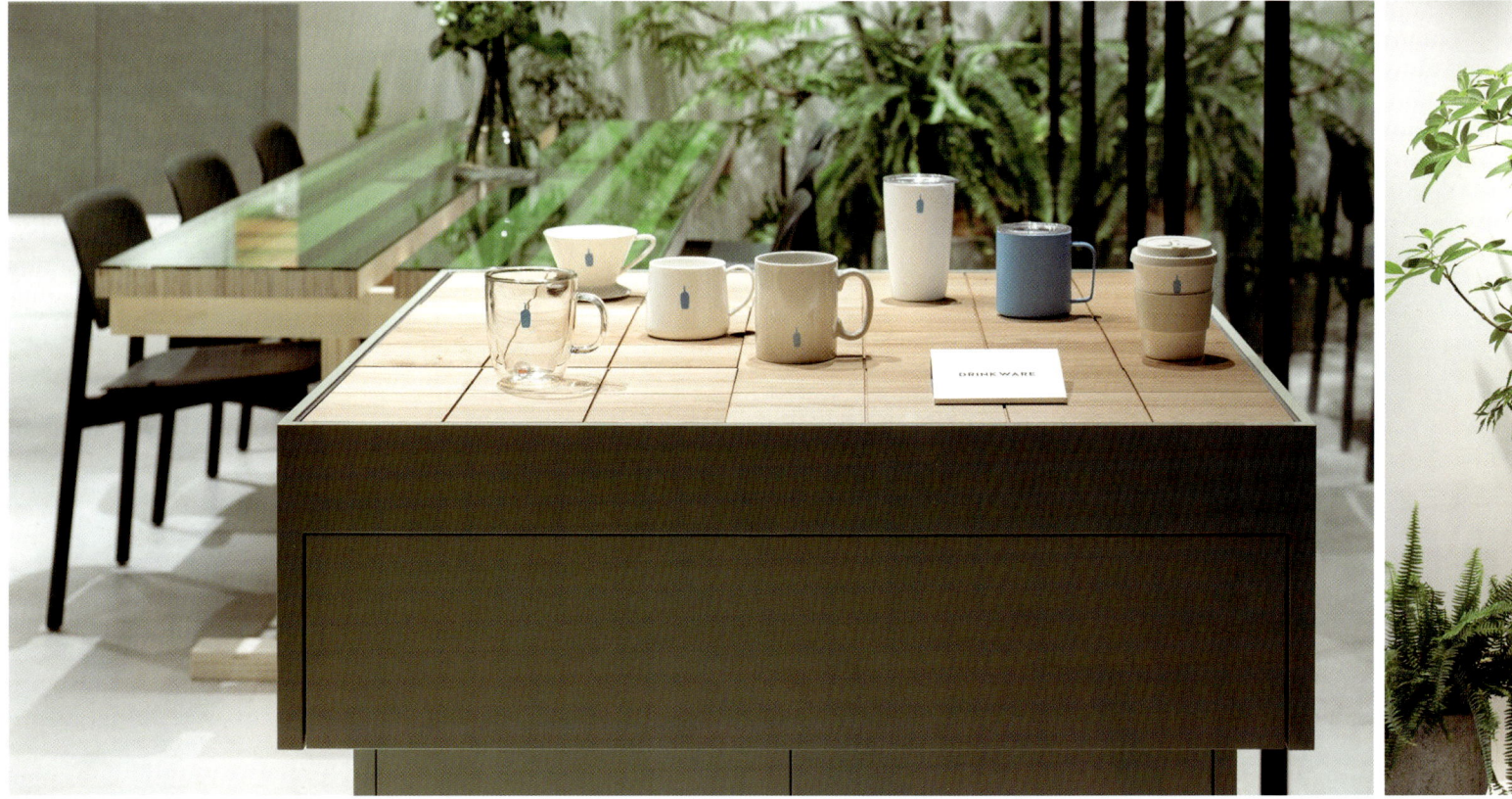

073

©HYACC

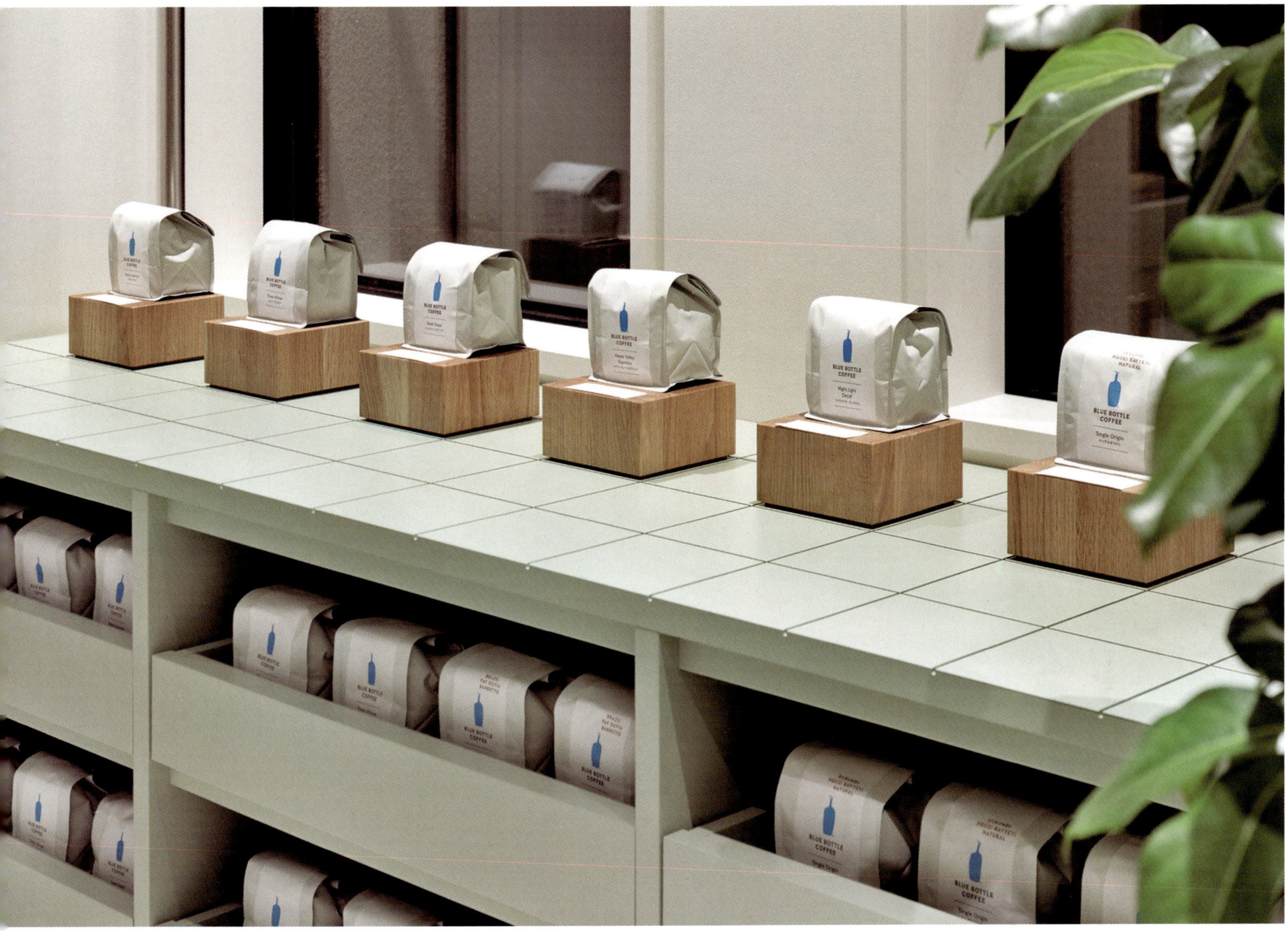

BLUE BOTTLE
COFFEE
Bella Donovan
ベラ・ドノヴァン
BLUE BOTTLE
COFFEE
Single Origin
シングルオリジン

BLUE BOTTLE
HOLIDAY YOKAN
WITH
CURRANTS, ORANGE
AND CINNAMON

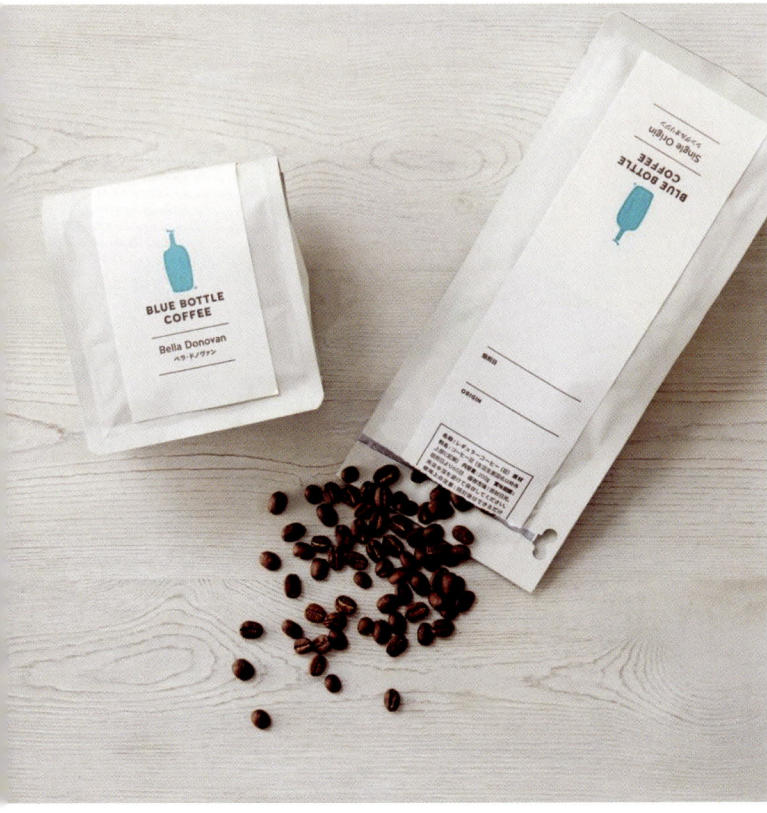

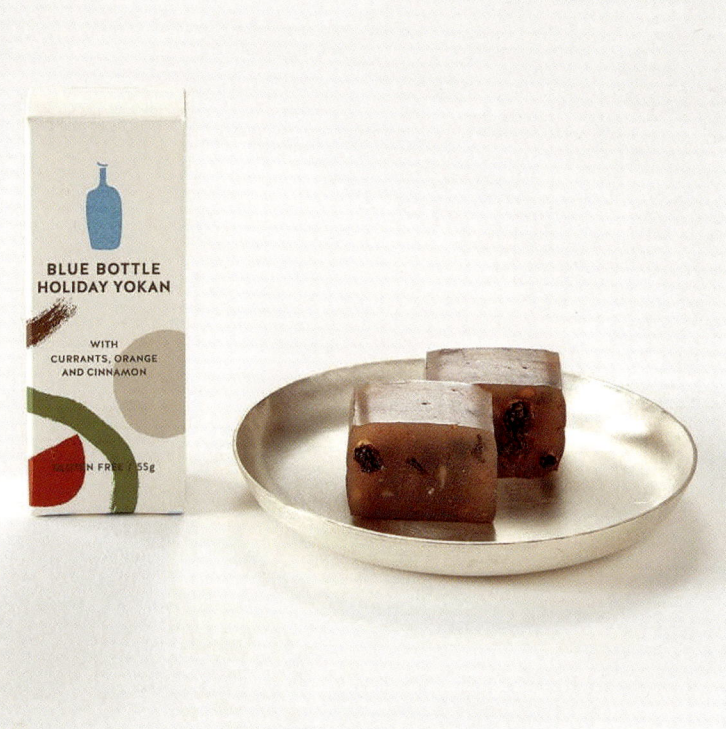

Kava & Chai coffee
Dubai, United Arab Emirates

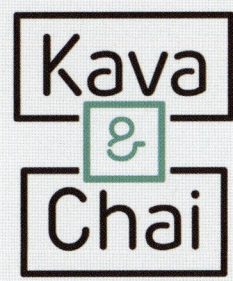

4 SPACE Design?

Originally established in Damascus in 2001, founders, Firas Alsahin and Amjad Hourieh, moved their practice to Dubai to be at the centre of this vibrant market. The emirate's booming growth in the commercial sector was an impetus for the firm to explore all the opportunities in the design industry. Overcoming an uphill battle, 4Space Design has gone on to create noteworthy projects in the UAE. Eschewing quantity for quality, profile of the project and relationship with clients, the studio credit its people's distinct ideas strategic business development.

Design: 4 SPACE Design   **Homepage:** https://4space.ae   **Founders:** Firas Alsahin, Amjad Hourieh
Location: Dubai, United Arab Emirates   **Photography:** Anas Al Rifai   **Area:** 170.00m²

Kava & Chai is a homegrown coffeehouse that projects Arab culture and traditions surrounding tea and coffee in a positive and intimate way. The client wanted the branding to exude the soul of the specialty coffee shop design. The aim is not just to document, but also to elevate the design. This 1,800 sq.ft specialty coffee shop design by 4SPACE is located in Dubai International Financial Center and has an exquisite interior which is unusual, astounding, and in complete balance. It has a striking concept and strong character that provides depth to the design.

Be startled on the ceiling that serves as the focal point of the design. The teal structured pipes with leaf graphics are inspired by the coffee roaster machine that the café uses at the back of the house. These massive pipes in the ceiling were designed using a parametric process which is based on the algorithm. It utilizes the latest techniques to set up the structural geometry of the project. The intention is to create a space where people can have a richer experience with a cup of coffee or tea without feeling time pass them by.

The color palette is originally based on the corporate brand of Kava & Chai and was enriched by the leaf pattern to promote its identity. The entire space provides a cool and calming attribute of sophistication, creativity, energy, and wisdom. Together with the customized lighting design that emits a soft glow, the natural quality of the materials creates a mesmerizing rhythmic visual effect enhancing the ambiance of the space. Terrazzo is mainly used on the tabletop because of its features being elegant and exuberant. It has a long-lasting finish that performs with a touch of class; valued for its integrity as much as its appearance. While on the floor, we have executed Palladiana terrazzo that is now known under sustainable products.

For the chairs and seating pods, 4SPACE used the color of the Pantone live coral that simply blends on the vibrant look and feel of the café. For the counter, we used Oakwood for its solid hardness with some inlays of coffee beans and tea leaf. 4SPACE have created a successful narrative from the brands' history and turned a dull and boring space into a more productive yet cozy and Instagram worthy place. Enjoying the mood amidst the busy world in the financial hub of Dubai.

About Us

At Kava & Chai we like to keep things simple, serving locally-roasted specialty coffee and premium tea in ways that everyone can enjoy. No bravado, no snobbery, no fancy brewing-gadgets: just great quality drinks and delicious light-bites at sensible prices. We were originally inspired by the first coffeehouses of the Middle East to create an environment where all are welcome. This inspiration led us to create the name 'Kava & Chai', a reference to 'coffee and tea' in the Arabic language. Order online or visit us today and rediscover the original coffeehouse concept: welcoming, easy-going and a place to unwind alone or meet with friends.

Kava & Chai

Our Inspiration

The first coffeehouses are believed to have originated in Syria before spreading to the Arabian Peninsula, then Turkey, by the 16th century. These coffeehouses became social hubs where people played boardgames, listened to stories and discussed the news of the day. Everyone was welcome. Differing views were exchanged, different cultures mingled, all whilst enjoying a freshly-prepared coffee. At Kava & Chai our philosophy is to continue the original coffeehouse theme. We pride ourselves on our coffee and tea quality, but make sure that everyone can enjoy them. We are the specialty coffee and tea brand without the unnecessary 'new-wave' complexity. Our staff are welcoming and we have added some delicious Arabia-inspired light-bites to complete the experience. Whether you are tired of style-over-substance expensive coffee shops or craving better coffee and tea quality than the standard global coffee chains, Kava & Chai is there for you.

#kavaandchai

قهوة البن

SLAB AND MEP SERVICES TO BE BLACK PAINTED FINISH (BP-01)
STEEL ROD SUPPORT
WALL PAINTED TURQUOISE PANTONE FINISH PT-01 (AS PER APPROVED CODE COLOR)
WOODEN CEILING WD-01 SUSPENDED FROM SLAB SUPPORT BLACK PAINT FINISH
SERVICE ZONE
WOODEN CEILING SUSPENDED FROM SLAB SUPPORT BLACK PAINT FINISH
EL±4.250M F.F.L
MDF ROUND PIPES IN TORQUOISE PATONE COATED - PT-01 FINISH (AS PER APPROVED SAMPLE AND CODE COLOR)
LED TUBE LIGHT SUSPENDED FROM SLAB
OAK WOOD CLADDING (W-01) (AS PER APPROVED SAMPLE)
LEASE LINE
EXISTING GLASS DOOR
EXISTING COLUMN
SERVICE STATION
OUTDOOR SEATING
EL±0.000M F.F.L
LOOSE FURNITURE LF-01 (CH-01)
SKIRTING TERRAZO TRZ-03 FINISH (AS PER APPROVED SAMPLE)
COUNTER OAK WOOD FINISH WD-01 (AS PER APPROVE SAMPLE)
ACCESS TO KITCHEN
SKIRTING TERRAZO FINISH TRZ-02

HANGING LED TUBE LIGHT
OAK WOOD CEILING (AS PER APPROVED SAMPLE)
AC DUCTING
AC DUCTING
STEEL ROD SUPPORT
SERVICE ZONE
SERVICE ZONE
EXISTING SLAB AND MEP SERVICES BLACK PAINTED FINISH (BP-01)
EL±4.250M F.F.L
MDF ROUND PIPES IN TORQUOISE PATONE COATED - PT-01 FINISH (AS PER APPROVED SAMPLE AND CODE COLOR)
MDF ROUND PIPES IN TORQUOISE PATONE COATED - PT-01 FINISH (AS PER APPROVED SAMPLE AND CODE COLOR)
BI-FOLDING DOOR BY LANDLORD
BI-FOLDING DOOR BY LANDLORD
EL±0.000M F.F.L
LOOSE FURNITURE LF-01 (CH-01)
SKIRTING TERRAZO TRZ-03 FINISH (AS PER APPROVED SAMPLE)
SMDB CABINET W-01
COUNTER OAK WOOD FINISH WD-01 (AS PER APPROVE SAMPLE)
LED SCREEN
UNIT DISPLAY
WOODEN PLATFORM WD-01 FINISH
UNIT DISPLAY

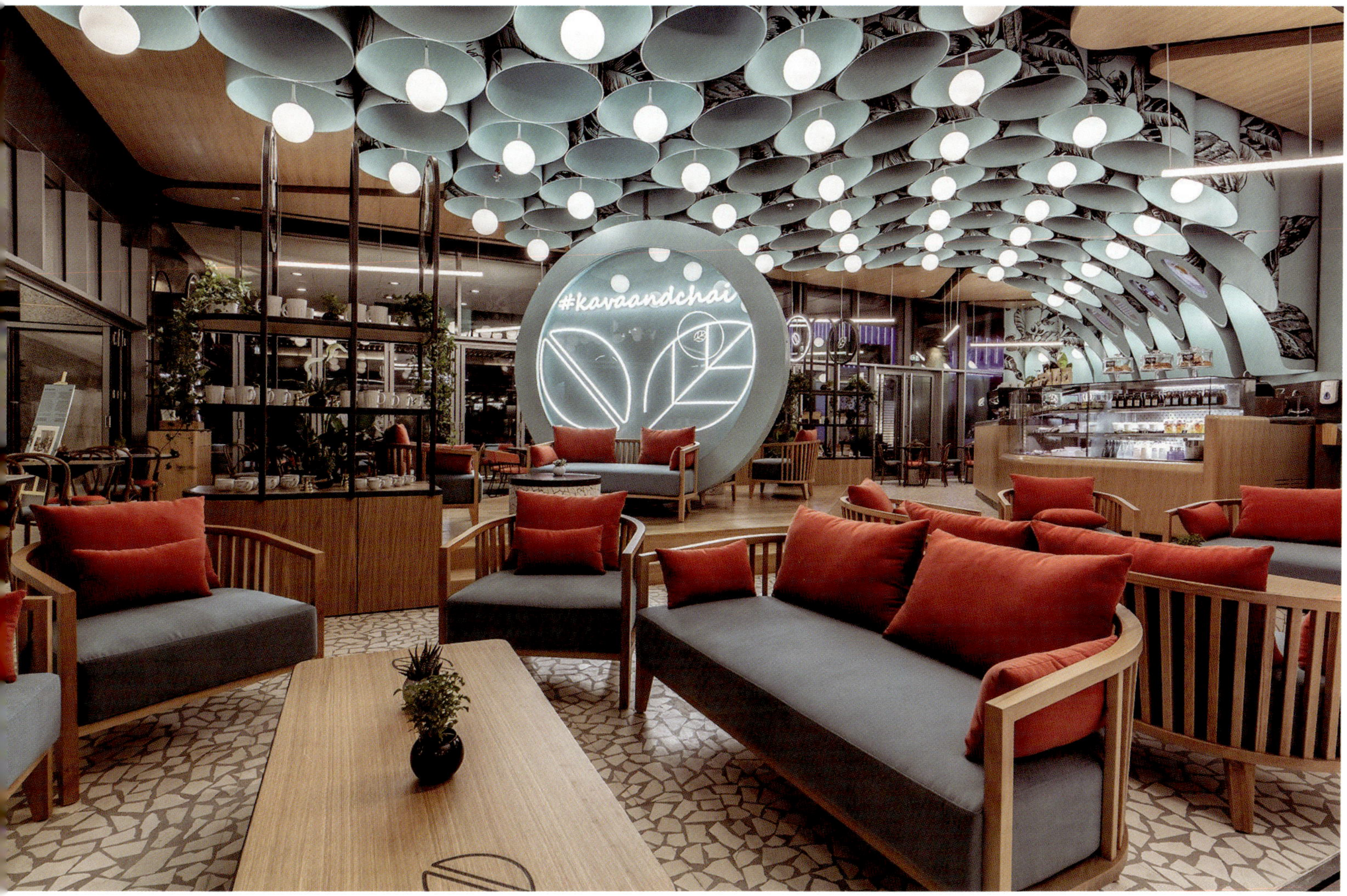

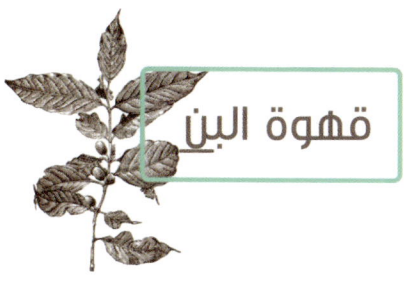

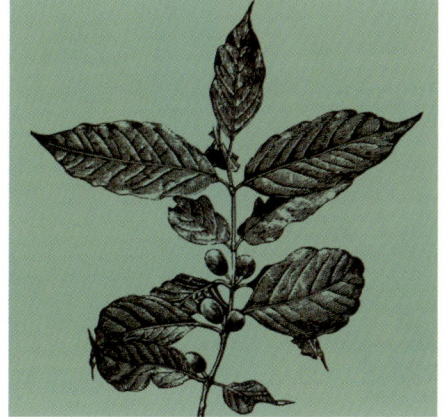

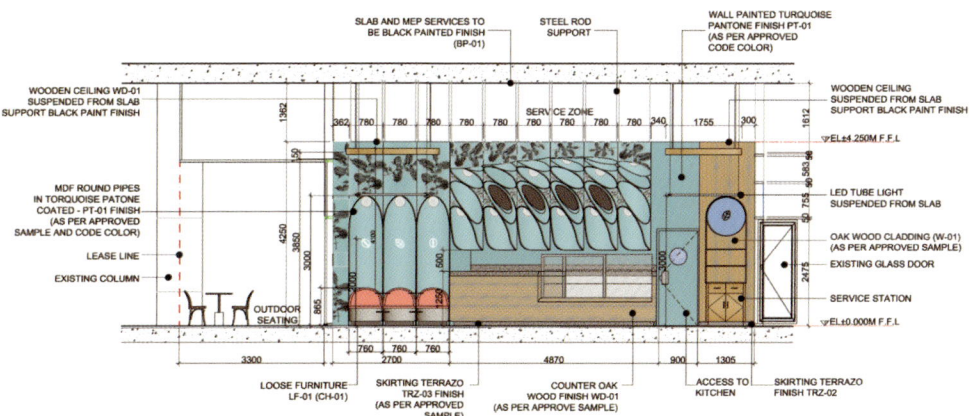

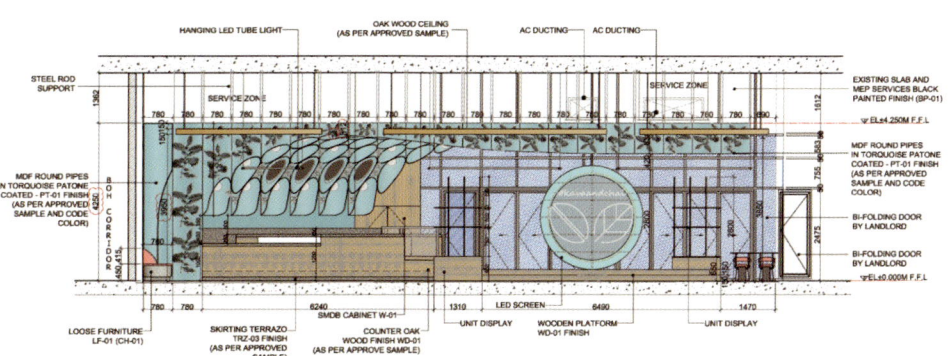

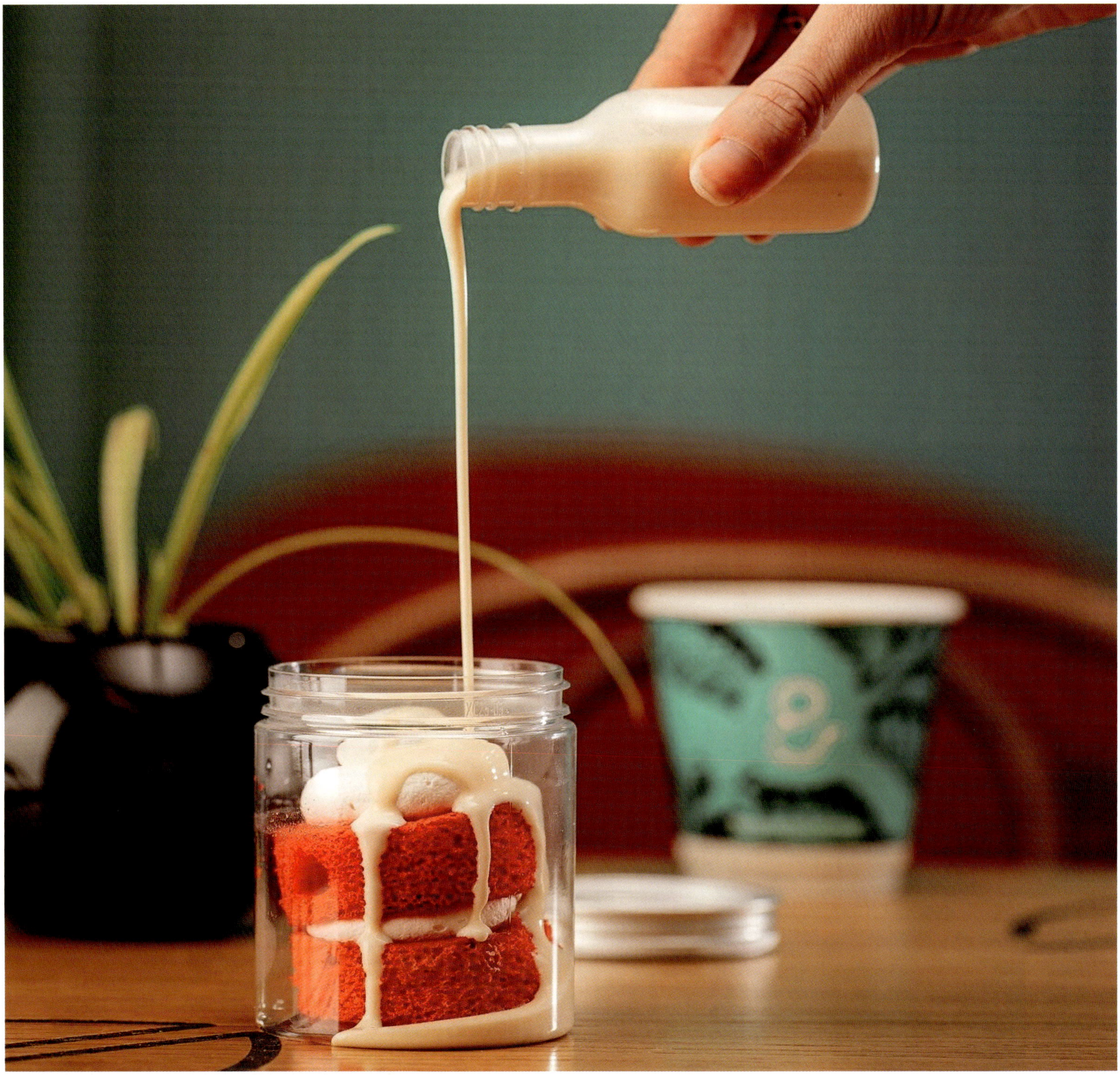

Kava & Chai
PALMICHAL
- Costa Rica -

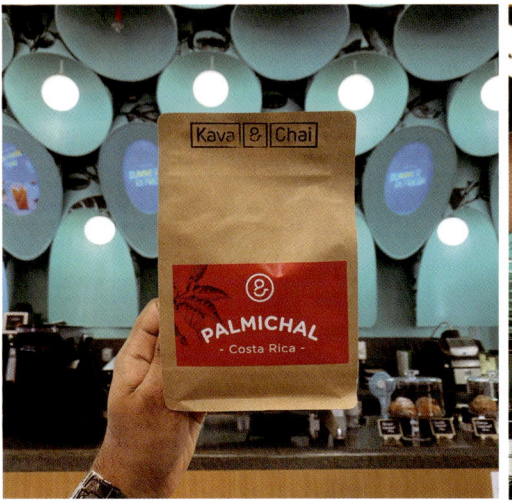

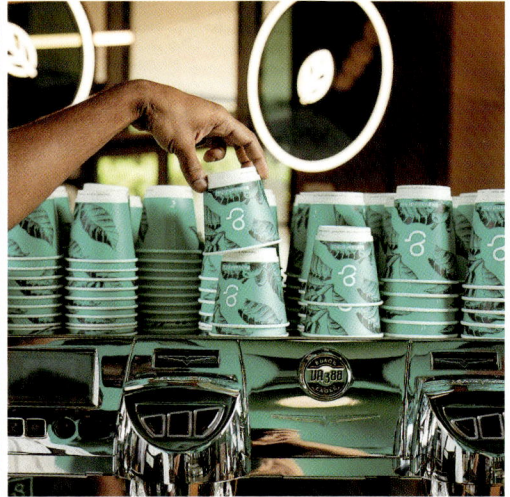

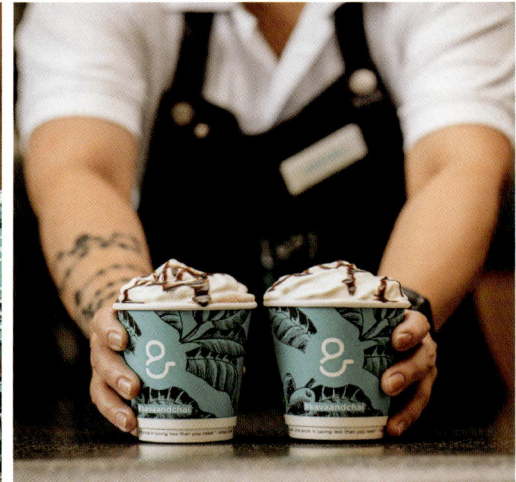

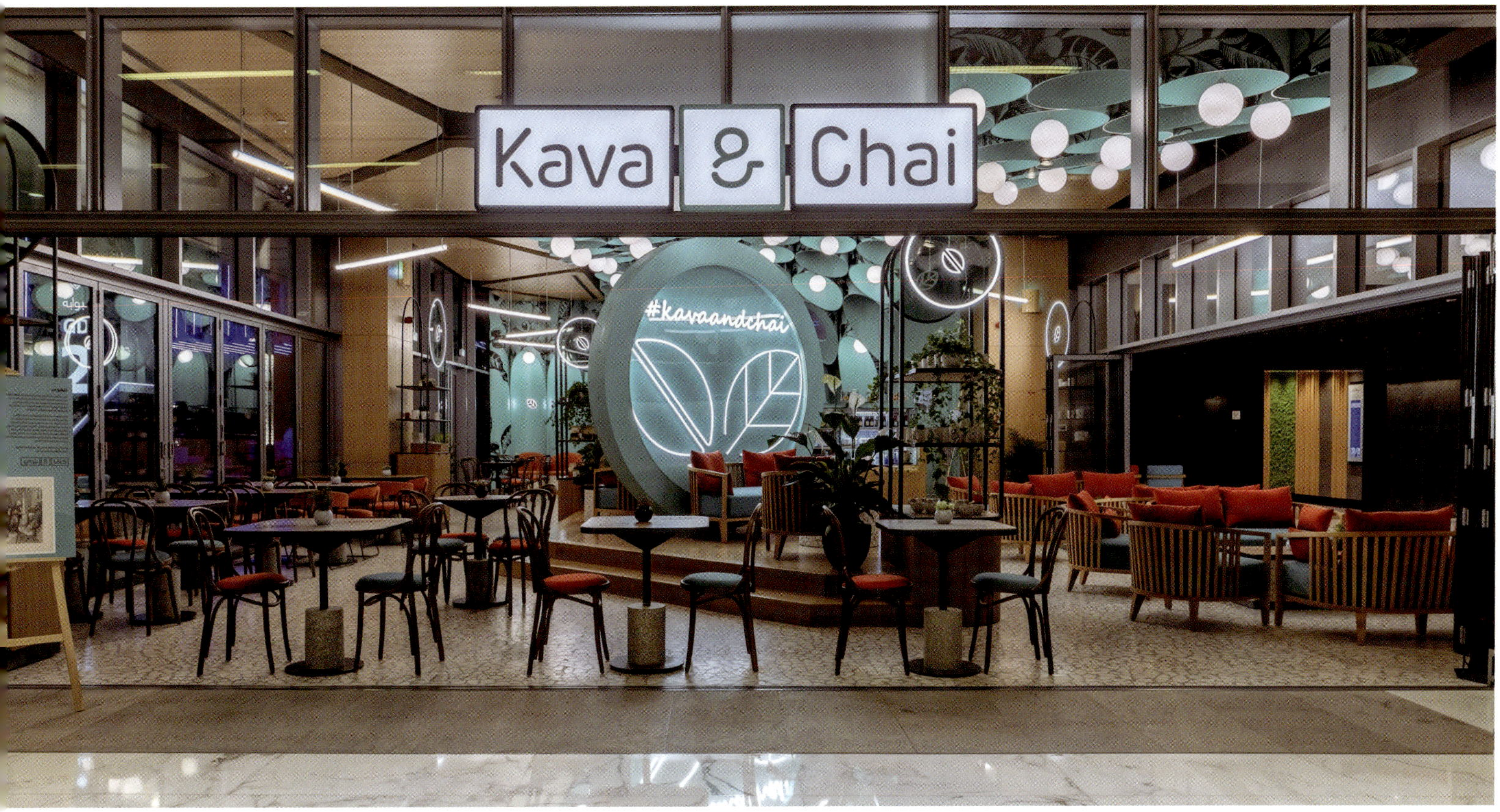

Kava & Chai
#kavaandchai

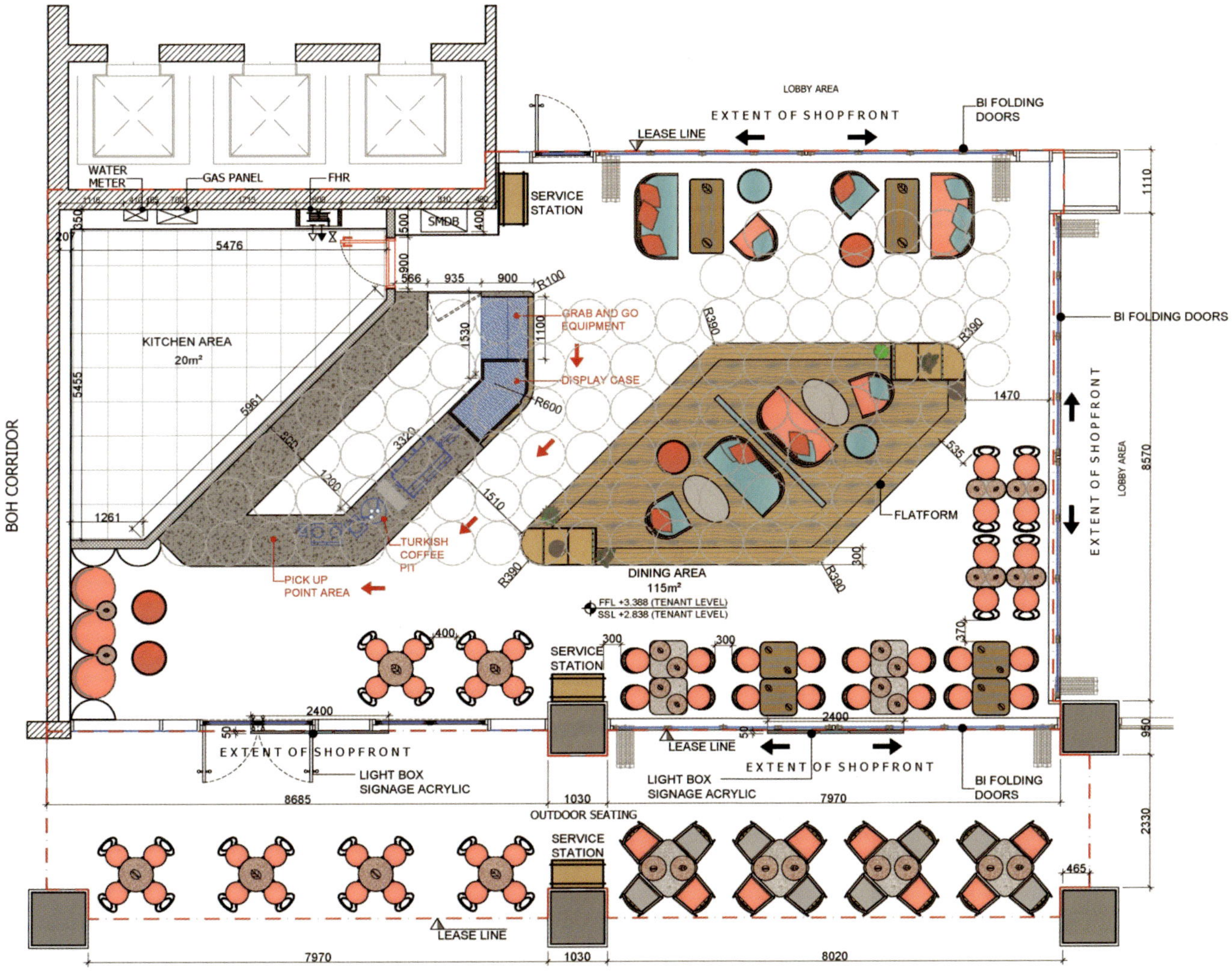

LOBBY AREA
EXTENT OF SHOPFRONT
LEASE LINE
BI FOLDING DOORS
WATER METER
GAS PANEL
FHR
SERVICE STATION
SMDB
KITCHEN AREA
20m²
GRAB AND GO EQUIPMENT
DISPLAY CASE
BOH CORRIDOR
TURKISH COFFEE PIT
PICK UP POINT AREA
FLATFORM
DINING AREA
115m²
FFL +3.388 (TENANT LEVEL)
SSL +2.838 (TENANT LEVEL)
LOBBY AREA
SERVICE STATION
LIGHT BOX SIGNAGE ACRYLIC
OUTDOOR SEATING

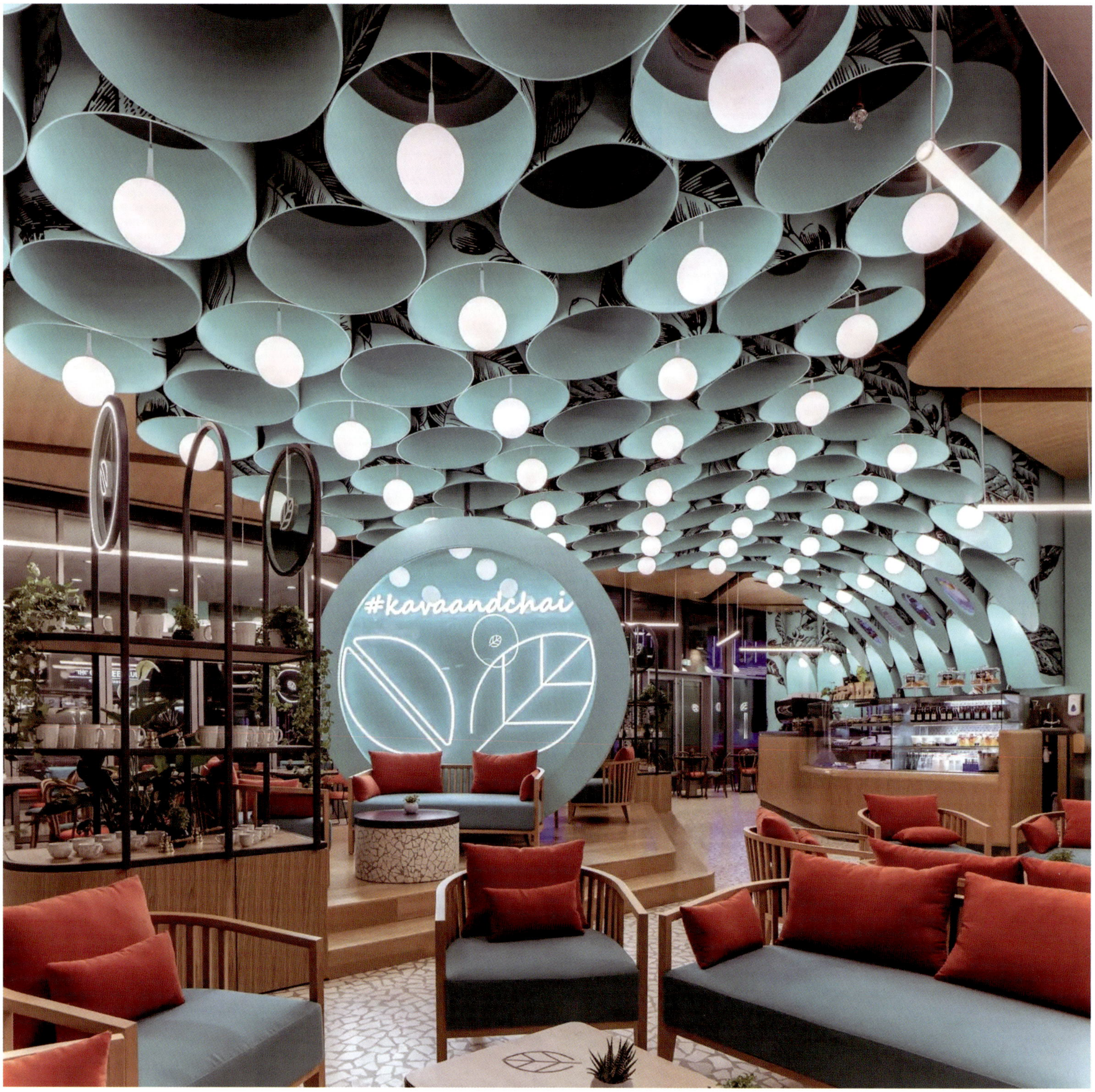
#kavaandchai

kavaandchai

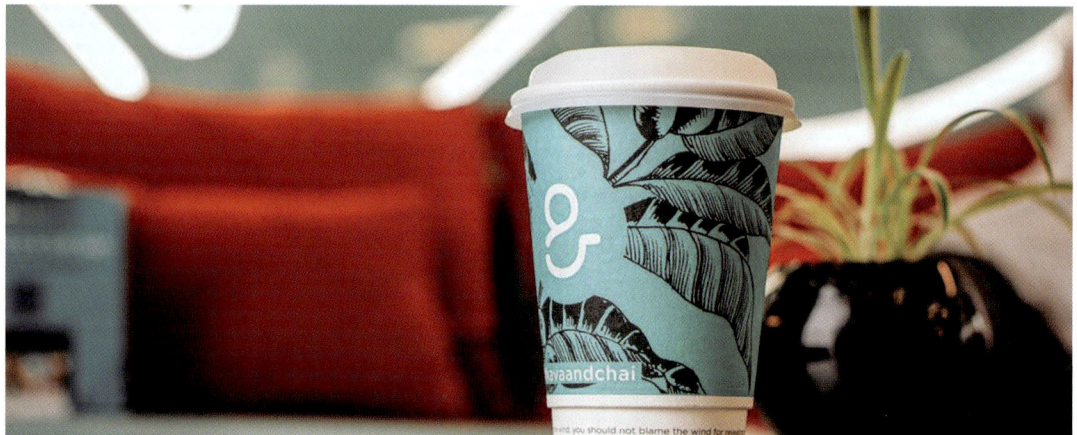

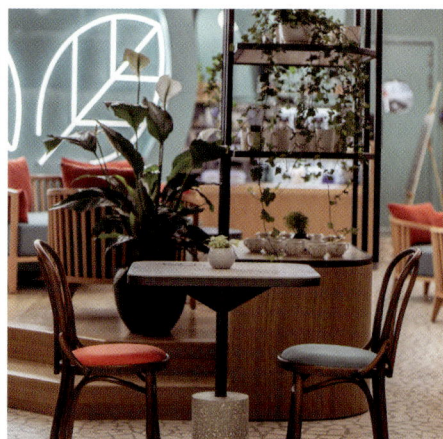

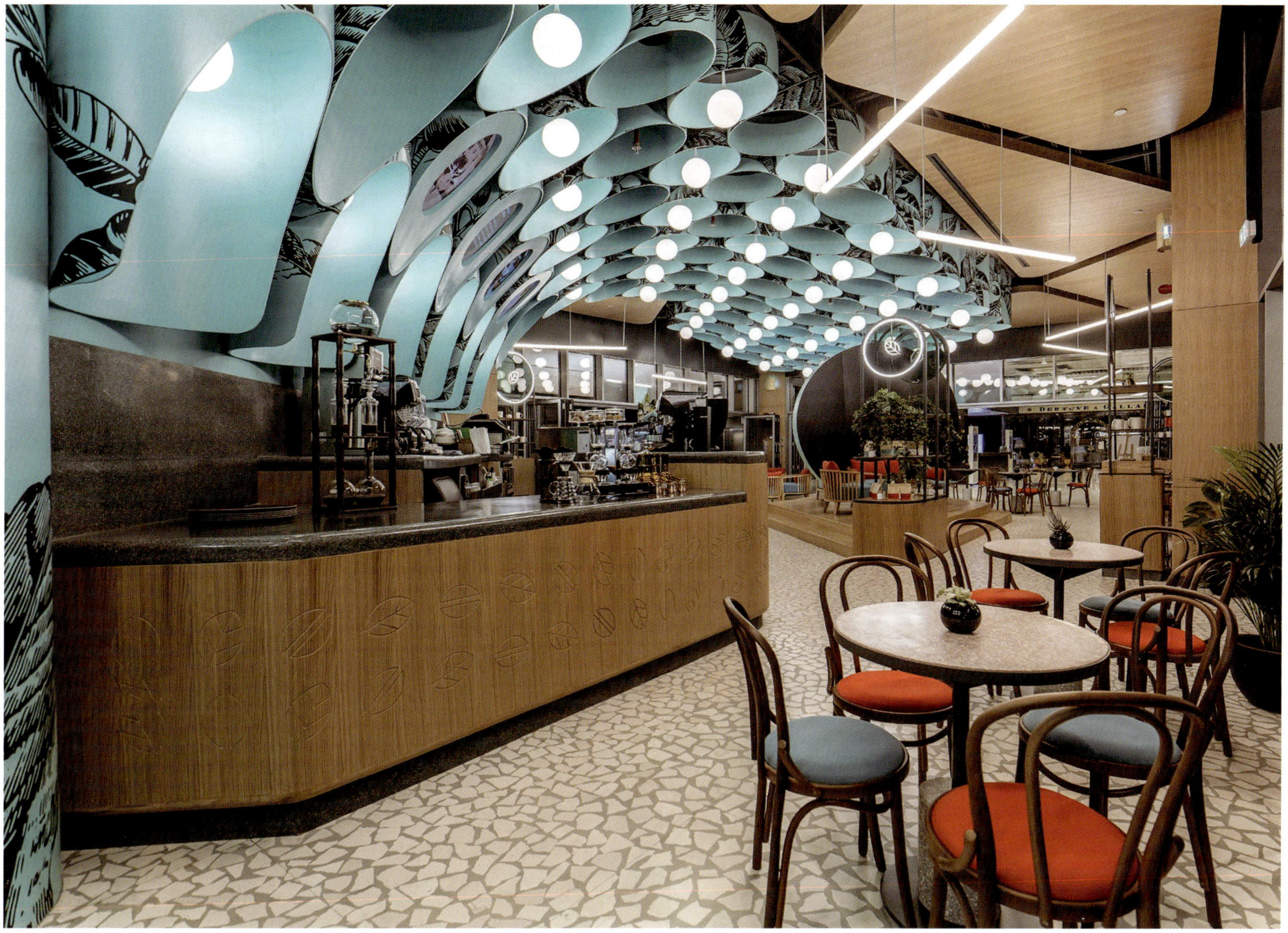

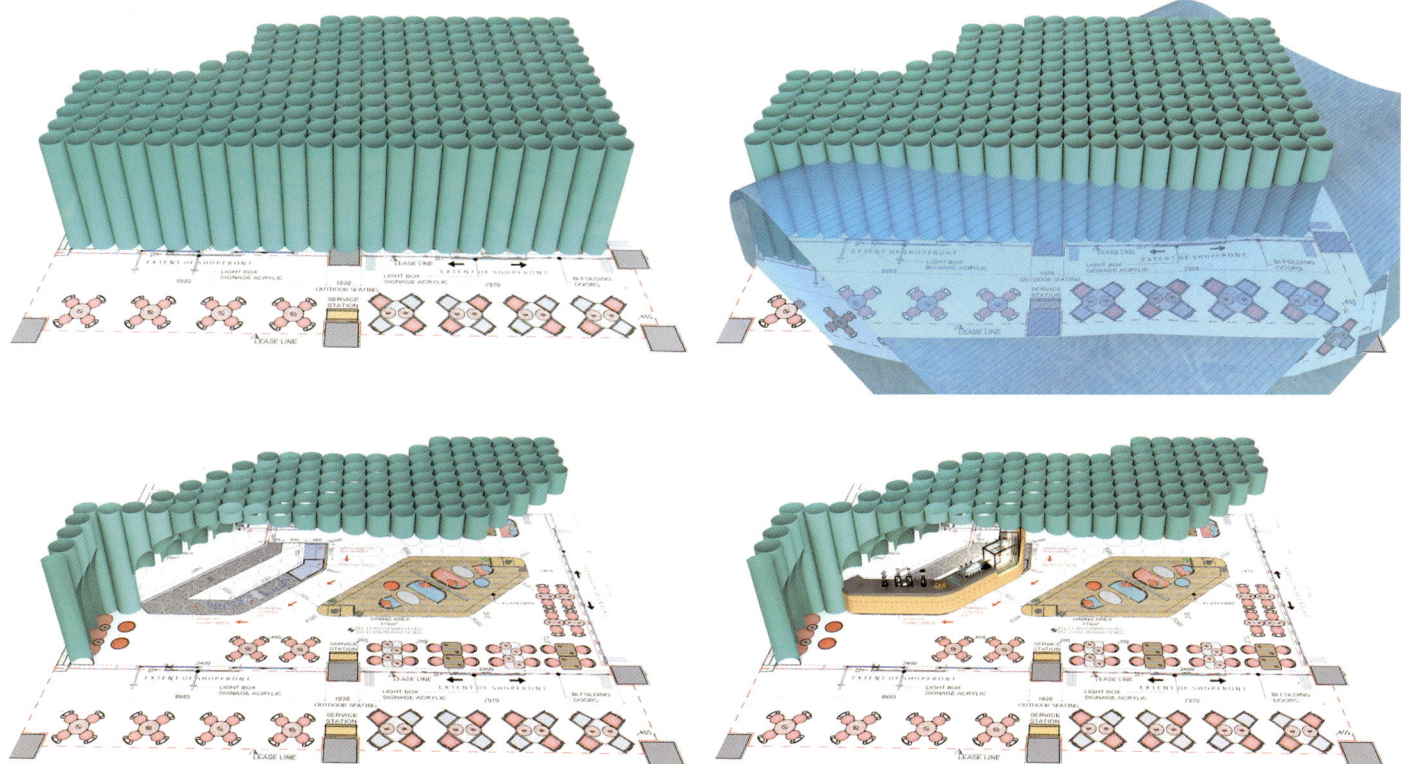

Kava & Chai

Specialty Coffee (The technical bit)

The Specialty Coffee Association (SCA) designed a process and scale that certified coffee tasters (Q-graders) use to grade Arabica (the best plant for coffee). If the coffee beans score 80 or more out of 100, they are classed as 'specialty'. We go through stringent processes to make sure all our coffees score greater than 80 and are officially classed as specialty coffee.

House of Eden
Melbourne, Australia

Ritz Ghougassian?

RITZ&GHOUGASSIAN established their architecture and interior design practice in 2015. Located in Melbourne, they focus primarily on the user experience and ones relationship to the built environment through an interiorised lens. Working across both public and private sectors a collaborative process ensures their work is considered and resolved. Architecture, interior design and landscaping intersect through a cohesive design language. Materiality is reimagined both internally and externally for each project, creating a palette that is coherent throughout the whole design. This process underlines a reductionist philosophy aimed at expressing the architectural volume through a simplification of the elements.

Design: Ritz Ghougassian **Homepage:** https://partyspacedesign.com **Print Collateral:** Hungry Worshop
Location: Melbourne, Australia **Photography:** Tom Blachford, Foliolio **Area:** 140m²

Ritz & Ghougassian reference brickwork for red-toned interiors of Bentwood cafe in Melbourne. Australian practice Ritz & Ghougassian used the worn red brick facade of this cafe in Melbourne as the reference point for the design of its interiors. The Bentwood cafe is set within the suburb of Fitzroy, an area of Melbourne well-known for its eclectic selection of eateries and bars. During the 20th century the building had been home to wooden furniture manufacturers C F Rojo & Sons, before becoming showroom for German brand Thonet. It now belongs to local coffee connoisseur Julien Moussi, who approached Ritz & Ghougassian to transform the 190 square-metre premises into a "unique space that pushed what a cafe should look and feel like". Set with this largely open brief, the architects decided to reference the colour and material palette of the building's exterior in the decor of its internal spaces.

"There wasn't a whole lot that was left other than the original red brick facade. However, I think the essence of its past are still there," Jean-Paul Ghougassian, director of the practice, told Dezeen. One of the cafe's peripheral walls has been clad with panels of russet-hued steel, some of which are perforated to allow glimpses into the kitchen.

Sheets of steel primed with red oxide have then been used to form a deep gridded ceiling, which the architects hope will create "pockets of light and shade, volume, and intimacy".

"A combination of prototyping and working closely with engineers and our contractor to create an effortless, seemingly floating ceiling span," explained Ghougassian. Warm timber chairs, tables, and shelving units also appear throughout the cafe in reference to the building's former occupants. Clay-coloured leather cushions have been used to dress the bench seats.

While most of the brick walls have been left in their found state, a handful of chunky concrete columns have been inserted as a subtle visual nod to Fitzroy's industrial heritage. Australian design studio Biasol also applied terracotta hues to a restaurant in Melbourne's South Yarra neighbourhood, whose aesthetic was inspired by the "evocative earthiness" of Middle Eastern architecture.

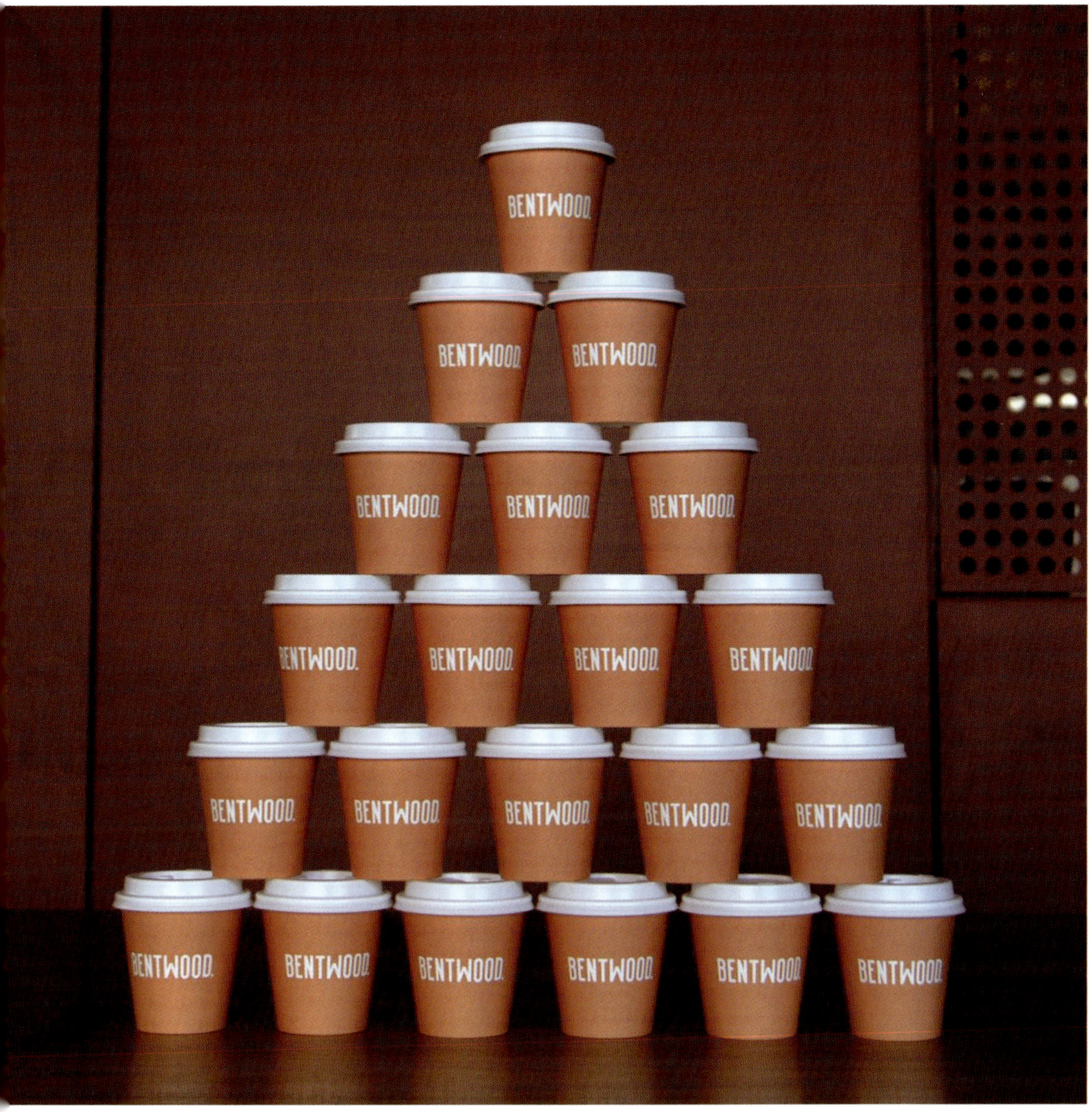
BENTWOOD.
BENTWOOD.
BENTWOOD.
BENTWOOD.
BENTWOOD.

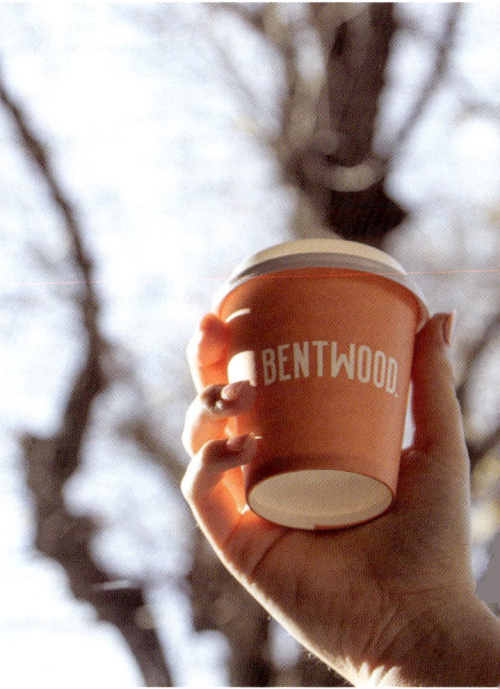

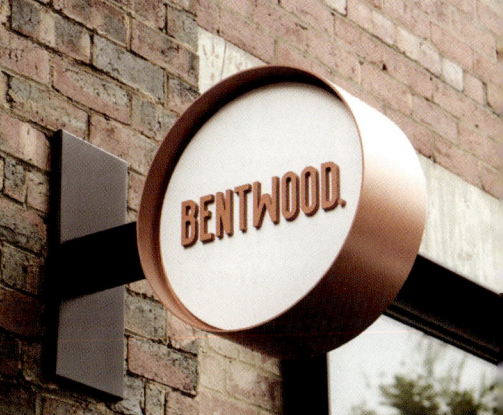

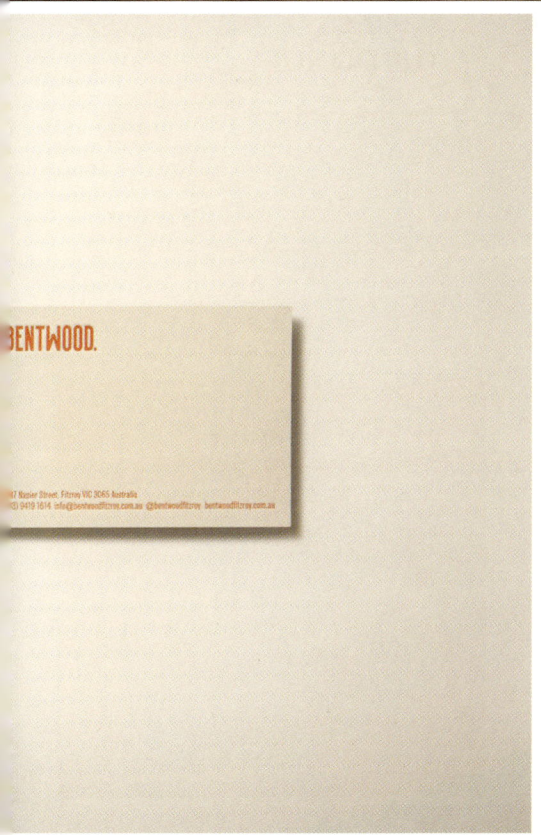

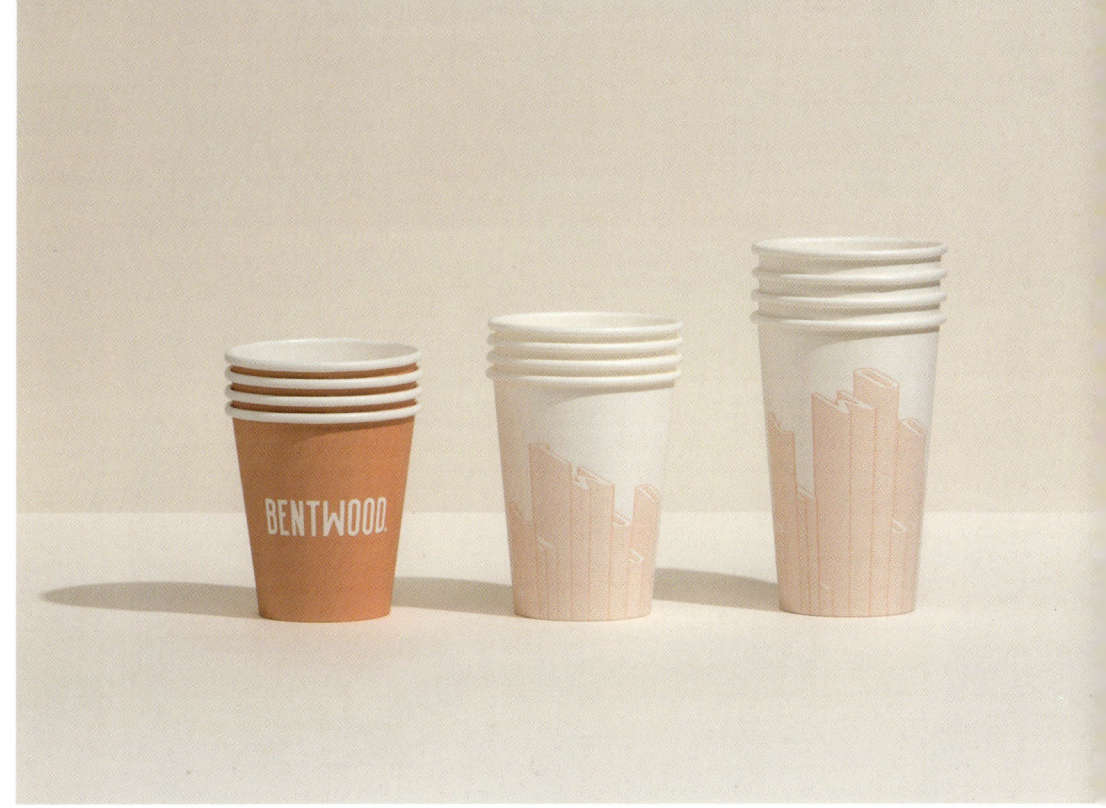

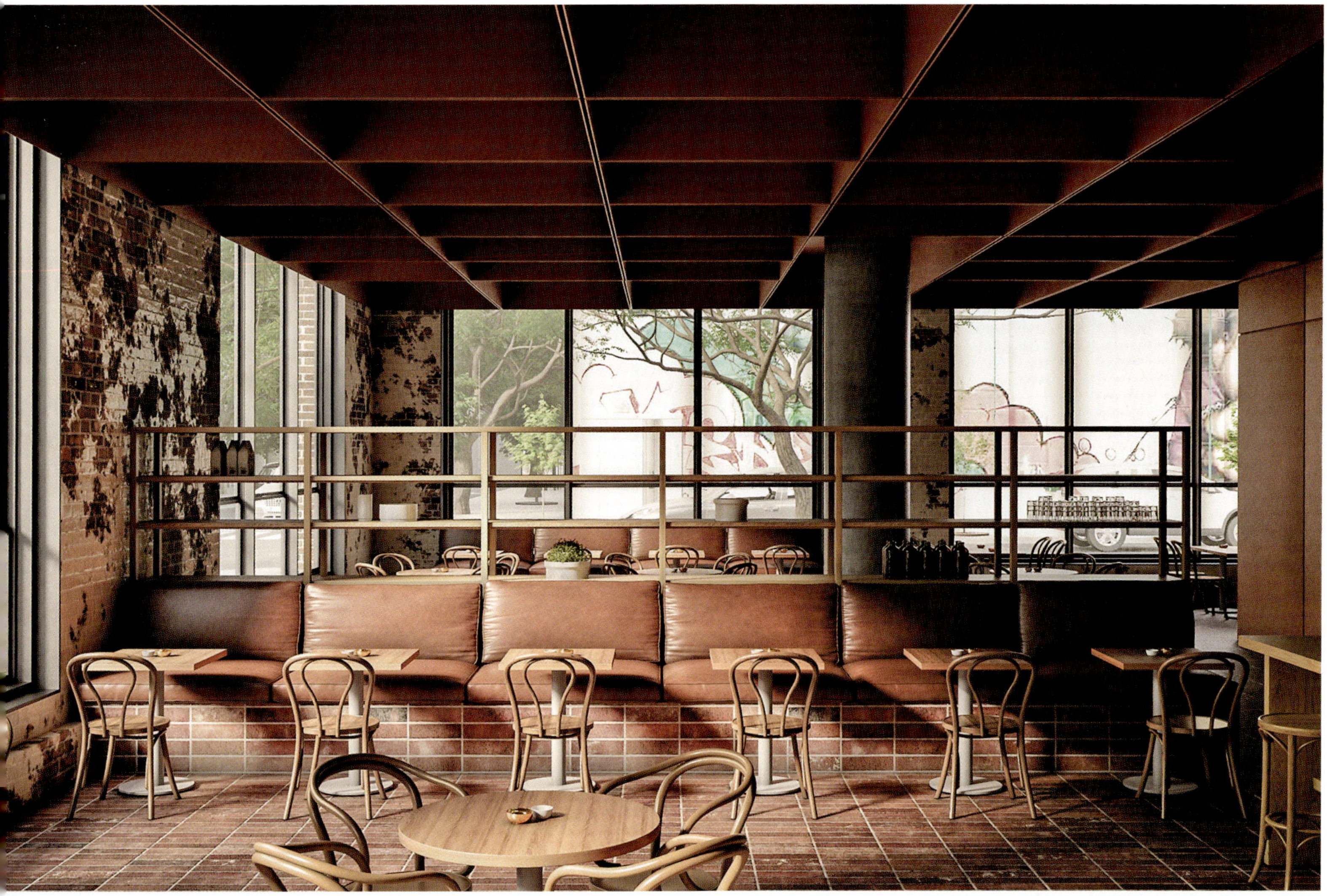

BOTTOMLESS
BRUNCH. $60
ESPRESSO MARTINI
BLOODY MARY
APEROL SPRITZ
MIMOSA

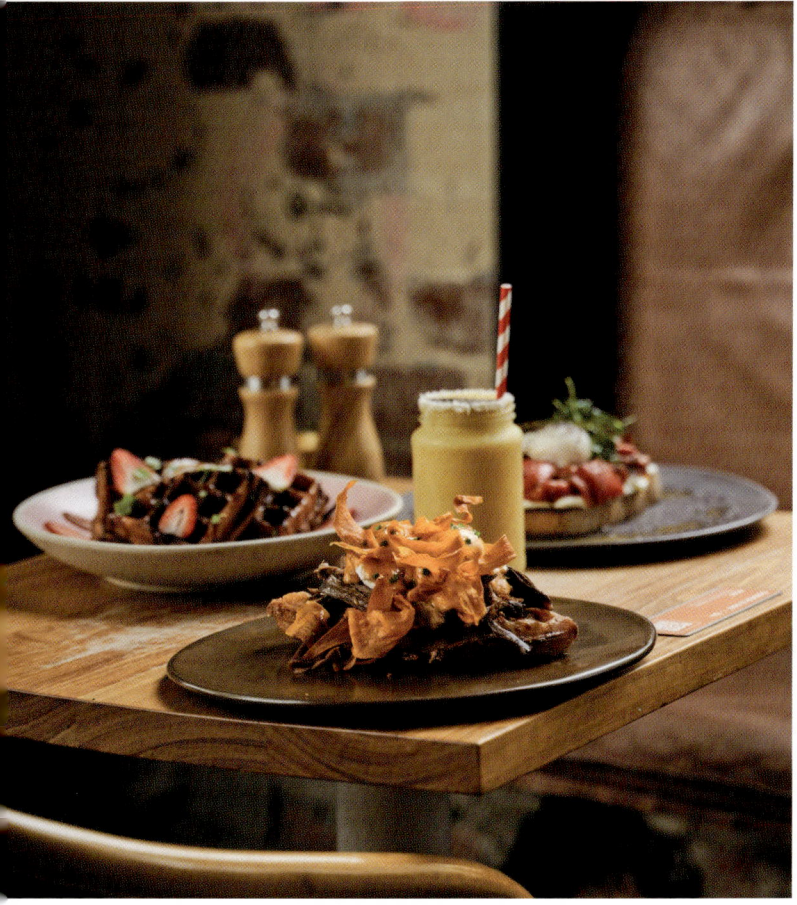

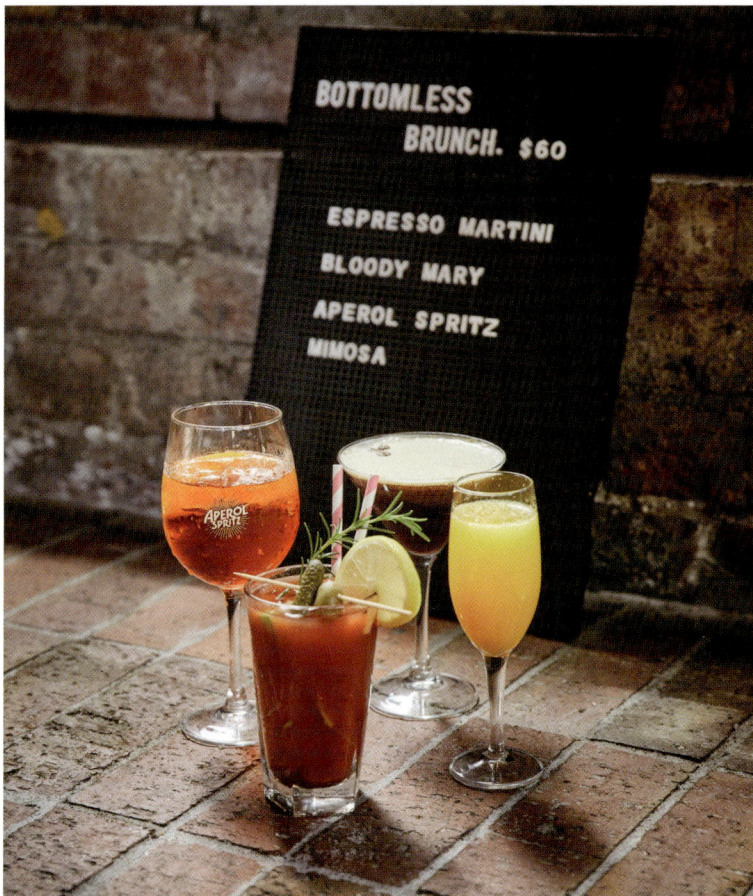

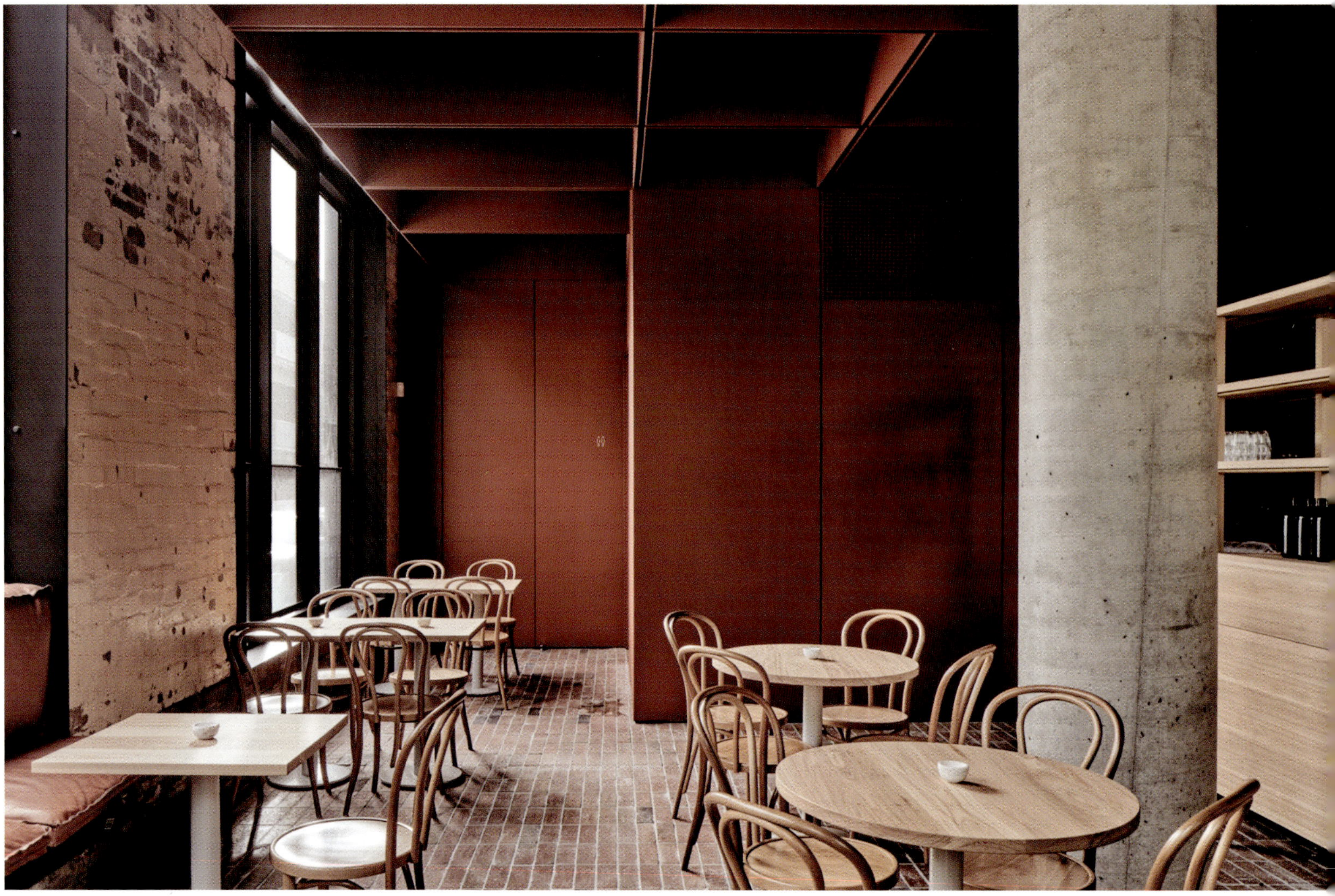

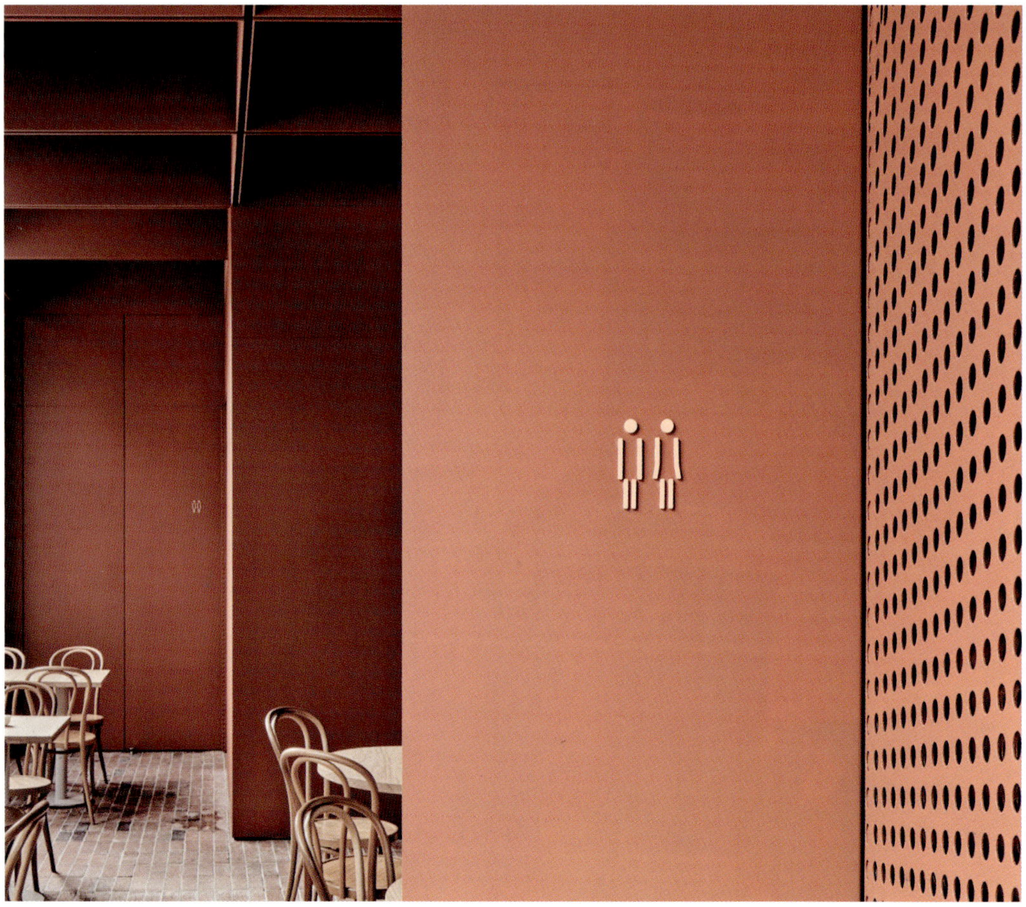

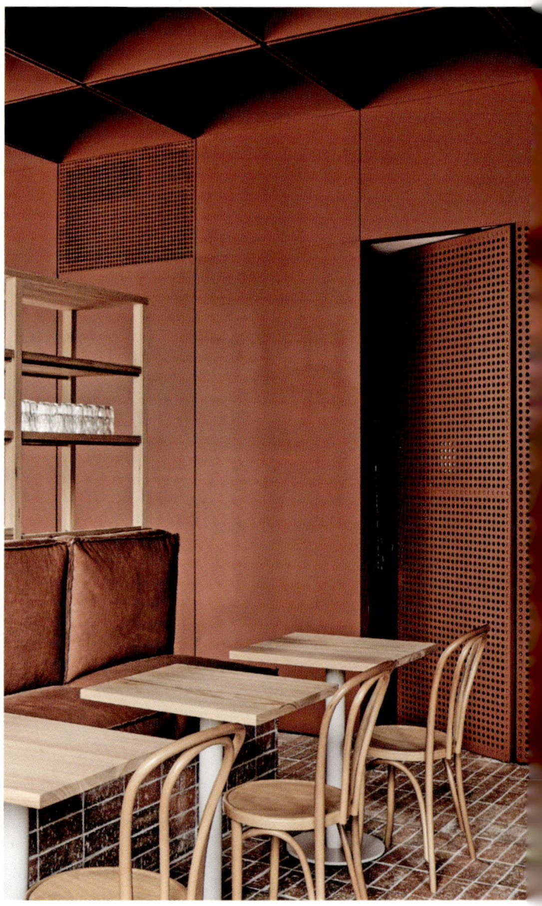

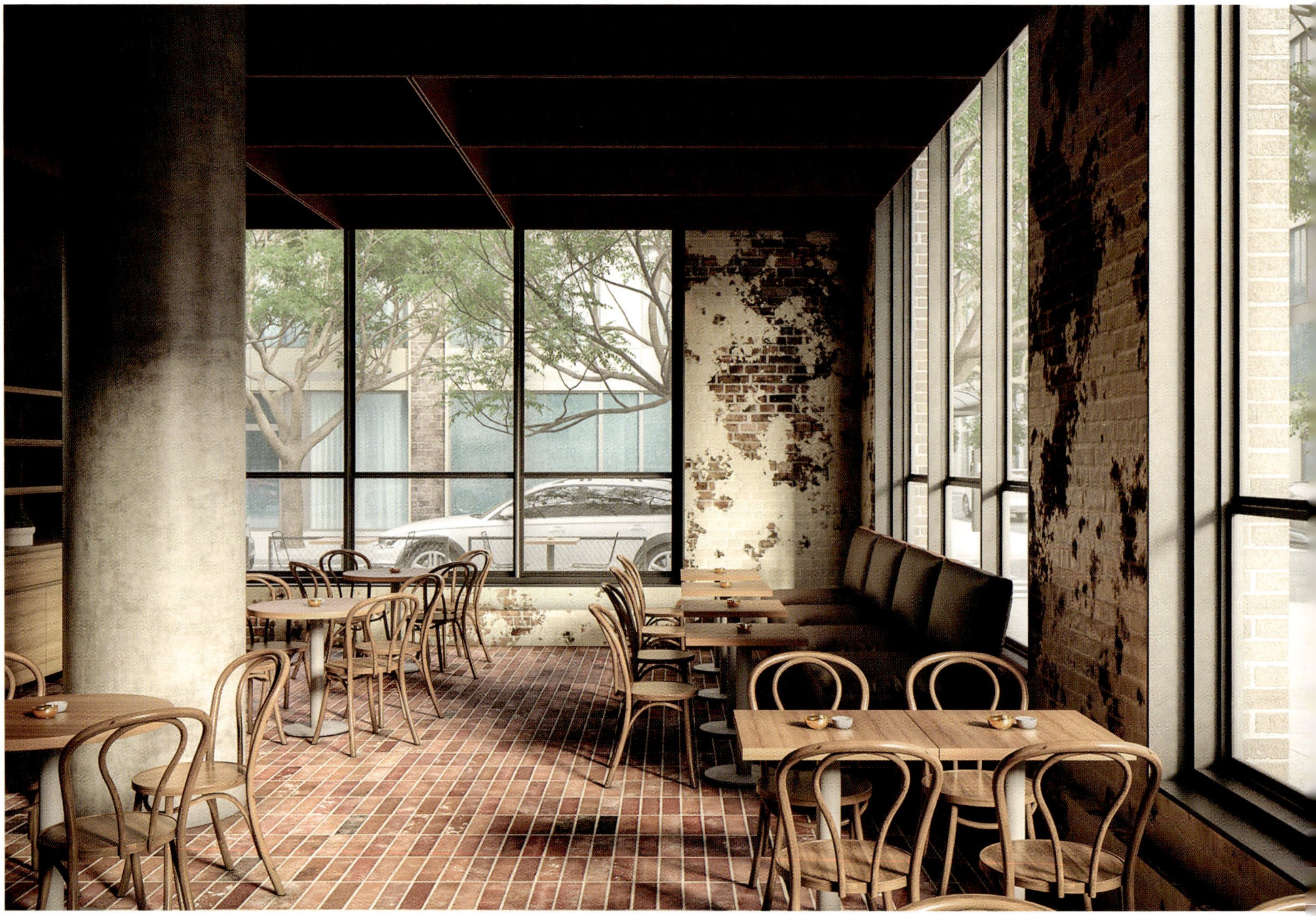

TH OUGH METICULOUS CRAFT AND P OCESS, NEVER HAS A CREATIO BEEN

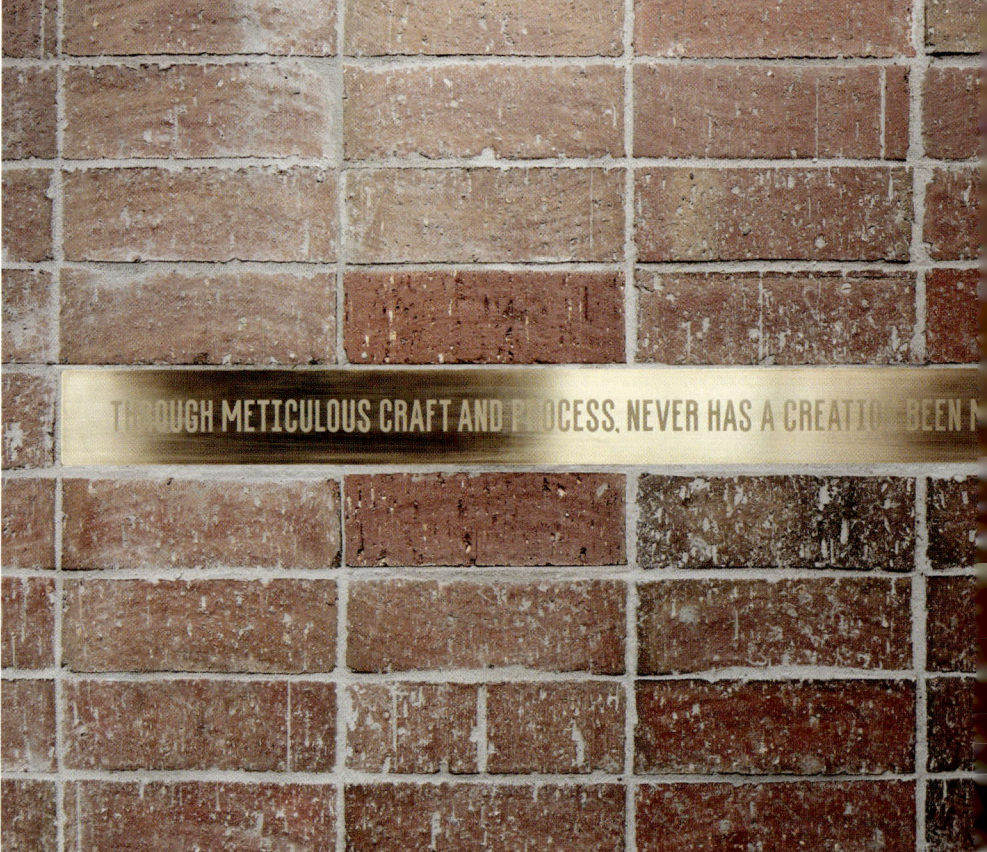

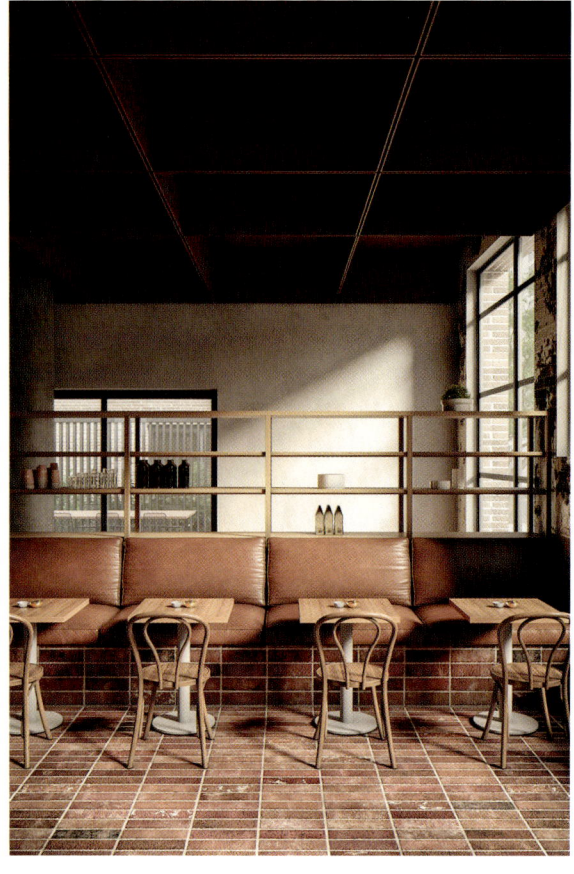

Bogen Bistro & Cafe
Bolzano, Italy

BOGEN
bistro

noa* network of architecture?

noa* is an award-winning architecture and design studio founded in 2011 by Lukas Rungger and Stefan Rier based in Bolzano (Italy) and Berlin (Germany). noa* stands for the essential expression of a collaborative work ethic: the young team of architects, designers and artists relies on an interdisciplinary design methodology that is constantly changing depending on the requirements of the respective project. The concept of "emergence", where the whole is far more than the sum of the individual parts, becomes the central strategy of a holistic approach to every design conceived by noa*.

Design: noa* network of architecture **Homepage:** http://www.danbrunn.com **Designer:** Dan Brunn
Location: Bolzano, Italy **Photography:** Alex Filz **Area:** 188m²

An ancient barrel-vaulted workshop hides under the dust of history along one of the oldest trade streets of Bolzano. noa* breathes new life into this space, transforming it into a welcoming bistro poised between historical heritage and contemporary finesse.

Bolzano's mercantile past echoes through the arcades of Via Portici, the city's main axis, which has been a trading hub for Italian- and German-speaking merchants since the 13th century. The goods that would later reach the whole of Europe were stored in its warehouses. Like Via Portici, also the northern parallel street, Via Dr. Streiter, has preserved much of its original appearance: today, it still passes through three medieval stone arches. It was mentioned for the first time in a document from 1498 and overlaps the old northern moat of the first town centre. About halfway down the street is a house that is hard to miss: with only two storeys, it is one of the lowest in the neighbourhood. An external staircase with an open corridor and round-arched portals provide access to the east side of the building, breaking up the compact street front. This house has been the backdrop for noa*'s latest interior design project.

The project involved the space on the ground floor where shoemakers, carpenters, carters, wood and fruit merchants worked in the 19th century, and where later the first restaurant on the street was established. Although the building preserved its charming original architecture, it had deteriorated over time. The Mayr family, current owners of the building, entrusted noa* with the restoration and interior planning for a new destination: the Bogen bistro.

Historical Heritage. The strong relationship with history was crucial in the definition of the project: both because the house is under monumental protection and because the design team wanted to emphasise to the fullest the original architecture of the arches, to which the bistro itself pays homage with the name "Bogen", German for "arch". On the exterior façade, the intervention consisted of a careful replastering in smoky white and an enlargement of the entrance arch. Here, a new tripartite black metal window follows the segmental arch and allows good natural lighting while providing an essential and timeless design.

For the interior, the underlying idea was to emphasise the four arches, which on both sides rhythmically mark the almost 19-metre depth of the room. To do so, noa* worked on both the horizontal and vertical dimensions. In the first case, the existing internal height difference was resolved with an oak platform at the entrance, while a grey-beige polished screed was chosen for the floor. In this way, there is no strong colour contrast with the walls, and the harmony of the shades enhances the whole space. On the other hand, noa* has designed the lighting so that the spotlights gently emphasise the curves of the arches. Except for the two tables at the end of the room, there are no pendant lights; floor lamps provide additional lighting.

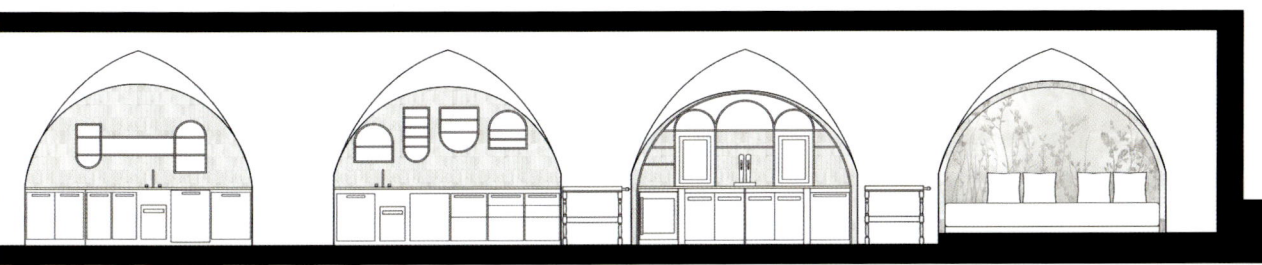

Longitudinal Section

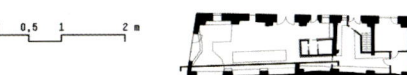

The long table as a convivial and informal solution in gastronomic ambience is a recurrent motif for noa*; in this case, the counter is also a worktop on the right side, without any stools and housing technical compartments. Other interesting details make this piece of furniture unique: the six legs are one different from the other and suggest an improvised table that a family might have made for itself. A mirror covers the central base and makes it disappear into the room. The top is a slab of Nacarado stone, chosen for its distinct veining and warm colour. Above the table, Roswitha's personal creation is the large floral composition that seems to pour from the ceiling. The hanging rattan lamps, which also recall the basket motif, find their place among the flowers.

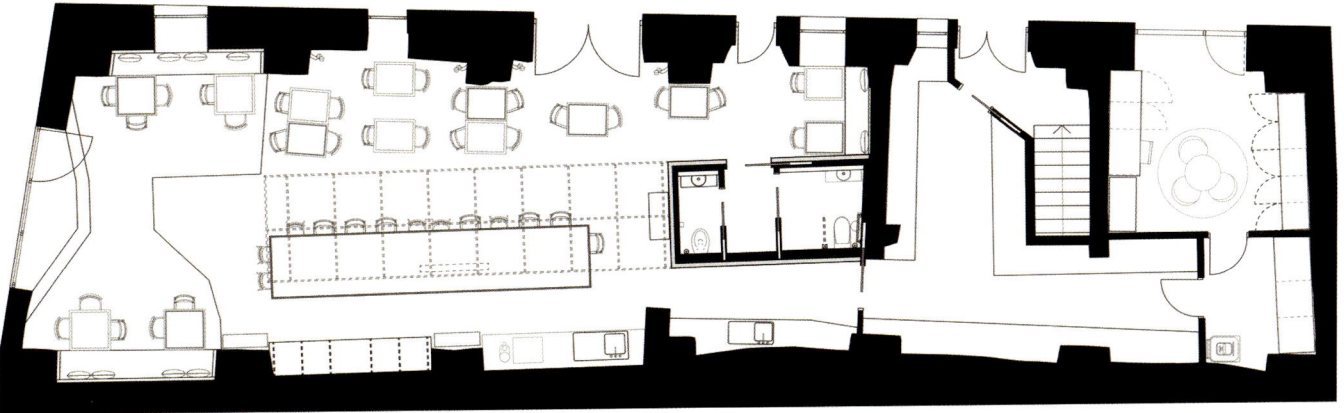

Bohemian Atmosphere. During the first meetings between the clients and the interior designers, which focused on the ambience of the bistro, the clients' desire to have a romantic, bohemian-style atmosphere emerged. In addition, the owner of the house, Roswitha Mayr, wanted to give the space a personal touch with her handcraft and artistic talents in the form of floral compositions. noa * took up these ideas and structured the design around a pivotal element: a welcoming 7-metre long counter placed under a ceiling of flower baskets.

Floor Plan

Cozy Alcoves. The shared space of the large counter contrasts with the intimacy of the small tables on the left side of the bistro, sheltered by the arches and overlooking the alley. The feeling of privacy is further accentuated in the first pair of arches, with seating built into the recesses and walls covered in fabric with an elegant floral print. The niche closing off the room is also designed in the same way. In a constant dialogue between past and present, noa* chooses to alternate new seats in wood and fabric with newly lacquered vintage chairs.

There are two service areas: the kitchen, which has been completely renovated and is located at the end of the room, and the toilets. These have been accommodated in a box, clad with perforated metal panels, on which the same floral motif of the arches has been printed. In this way, noa* combines the technical requirements of acoustics with the venue's aesthetics: the insulation panels are not visible under the perforated metal surface.

SACHER
EIN LEBENSGEFÜHL

BWM Architekten und Partner?

BWM Architekten are a multi-national architecture office that operates throughout Europe. Their main areas of focus are architecture, interior design, culture and hospitality. Founded in 2004 and led by Erich Bernard, Daniela Walten, Johann Moser, Markus Kaplan and András Klopfer, the company and the 70-member team stand for a personal approach and a cooperative development process. Whether designing interior spaces, residential and urban construction projects, or museum and exhibition concepts, BWM Architekten always work out a specific project's unique formal language and the corresponding design concept in strategic workshops with the client.

Design: BWM Architekten und Partner **Homepage:** http://www.danbrunn.com **Team:** Erich Bernard, Aleš Košak, Ismail Berkel.....
Location: Vienna, Austria **Photography:** BMW Architects/ Severin Wurnig **Area:** 70 m²

BWM Architekten have tastefully inserted elements from the 1920s and 1950s into the newly designed Salon Sacher and have created a cohesive unit that marries the past and the present. Following the redesign of the Sacher Eck at the end of 2017, BWM Architekten have now also redesigned the former Sacher Stube, now known as the Salon Sacher. The black and coral colour scheme and the 1920s and 1950s elements congenially complement one other. Arched globe lights highlight the bar's role as a centrepiece, and tinted mirrors, black lines and metallic effects perfectly round off the overall look. The preserved stucco ceiling is an absolute eyecatcher: Previously hidden behind the dropped ceiling, this historic jewel was discovered in the course of the construction work. A large section of it is now displayed in all its splendour.

Space for history

Original elements form the main part of BWM Architekten's concept for the redesign of the Salon Sacher in Vienna's venerable Hotel Sacher. The eye is immediately drawn to the historical stucco ceiling dating back to the time the building was built. Previously hidden behind the dropped ceiling, it is now accentuated by means of a generous opening in the 5m-high ceiling. The tent-like structure of the room dissolves the usual spatial boundaries, creating an impression of surprising vastness... The central theme is also consistently reflected in the furnishings. The original Thonet chairs have been restored and reupholstered; the tables have partly been left in their original state, keeping their brass bases while being furnished with new stone tabletops.

Coral-coloured redesign

Coral-red accents add a fresh note and pay tribute to the era in which the former Sacher Stube originated. "Combining black with light colours and shiny effects is typical for the 1920s, but also for the 1950s, which have much in common with the 20s," BWM architect Erich Bernard explains. "The light colours that were popular at that time – besides the classic Art Deco creme colour – included pink, peach and coral, which has something of an exotic connotation as well. For us, coral red is also a modified form of the characteristic Sacher red, transported into today's world."

Black provides contrast

"Understated elegance in black and white in combination with rounded shapes, longitudinal oval mirror elements and eccentric pedestal lamps at the bar are the other main ingredients of the design concept," says BWM project manager Aleš Košak. The geometric lines and black-and-white pattern of the floor are inspired by a sketch by the famous architect Josef Hoffmann, who was also the co-founder of the Wiener Werkstätte. Black is also the colour of the vertical trim-edges on the walls and ceiling and the wainscot panelling. Together, all these elements create a contrast to the lightness of the coral-coloured upholstery.

For us, coral red is also a modified form of the characteristic Sacher red, transported into today's world. -Erich Bernard

"Coral-red accents bring a fresh tone to the room and are a nod to the time the original Sacher Stube dates from."

Crowning element for the centrally positioned bar

The dominant element of the interior design is the new, oval bar with a surface consisting of a stone slab set in brass and sides clad with bronze-coloured, facetted mirror strips – all emphasised by arched globe lights designed by the Viennese designer Megumi Ito. The crowning glory is the historical stucco ceiling tantalisingly peeking out of the oval opening; a neon border draws all eyes to it.

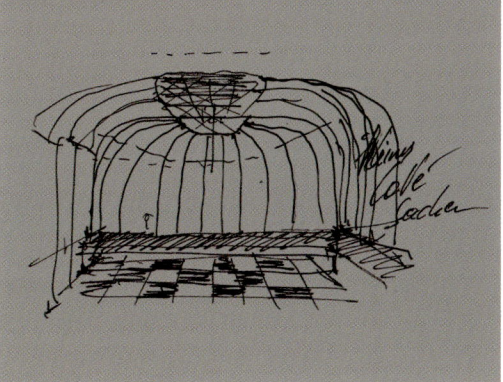

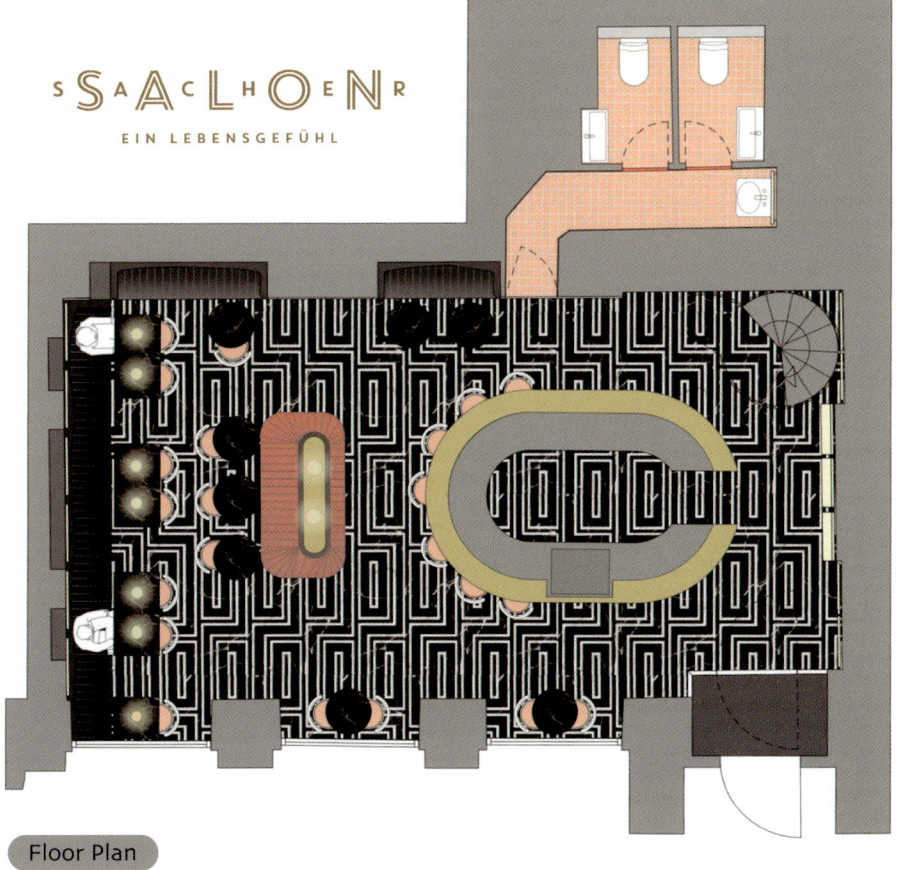

Floor Plan

Down to the last detail …

A central, coral-red settee divides the room into the bar area on the one hand, and the café zone with classic coffee house furnishings and black chairs on the other. The salon provides seating for 40 guests and is slightly larger than its predecessor. The design of the café's bathroom, located in a small annex with a ceiling height of 5½ m, is an amplification of the approaches applied in the elegant salon. Decorative wallpaper and large tinted mirrors create an illusionistic interior. Successful collaboration – in 2017 BWM redesigned the Sacher Eck

The Salon Sacher is not the first collaboration between BWM Architekten and Sacher. In 2017 the Vienna-based architecture firm already redesigned the Sacher Eck in the iconic Hotel Sacher. BWM Architekten based their redesign on traditional, imperial Viennese interiors, of which the iconic Hotel Sacher is a prime example. The combination of the classic Sacher colours – wine-red, gold and black – and the typical Viennese materials of velvet, brass, dark wood and black and white marble created a new identity. By extending the café and confectionery and breaking through the ceiling to the mezzanine, BWM skilfully drew attention to striking design elements such as the monumental chandelier. In addition to the confectionery, where the original Sacher products can be purchased, there is now also an additional spacious café area on the bel étage with a view of Kärntner Strasse and the Vienna State Opera.

Café Flor
Schiphol, Netherlands

CAFÉ FLOR

creneau Int?

We like to think before we do. That is why we start with the facts, then turn them around for you, offer you another take on things. Our job is to think harder. To see differently. And to treat you to fresh ideas. You have plans—a hotel, a restaurant, offices, a whole city block, or maybe just your product's packaging. And you have questions—from how to master the master planning to which chairs, maybe even what should be on the menu. Before we do any designing or building, we strategise. We'll sit you down, dig for your reasons why, understand your world and come up with a blueprint for your business. The numbers, yes, but all the feels, too. Because a good strategy is one that rhymes ROI with speaking to your customer's heart.

Design: creneau.com Int **Homepage:** www.creneau.com **Team:** Bart Canini, Serge Haelterman, Davy Grosemans.....
Location: Schiphol Airport, Netherlands **Image:** creneau.com Int **Area:** 70 m²

A place for The Ambitious City Girl: Schiphol airport had a very clear target audience in mind. Lots of color, a central bar with botanical graphics, while lattes and cocktails are stealing the show.

Café Flor is the name of this eclectic bistro that gets you in the holiday spirit even before you entered the airplane. This is not your ordinary airport bistro: it is the only full service restaurant at Schiphol, where it is evident you are served at your table. At Café Flor you can enjoy delicious cocktails or a healthy meal and finish with a Macchiato brewed by the barista on duty.

At Café Flor, all kinds of seating options are available for you to sink into: ottomans, sofas, benches, barstools, or you might prefer the piano stool with the accompanying piano? Yes, we introduced a piano in this space. All visitors are welcome to take a seat for their minute of fame.

The eye catcher of the room is the bar with graphic flower patterns on top. The playful element of the flowers, the colors and the detail of the tiles that continue from the floor onto the bar create a small moment of magic. Besides the interior, we also developed the entire branding for Café Flor. We opted for cheerful, natural colors. We worked these colors into the logo and the botanical, tropical flowers that we

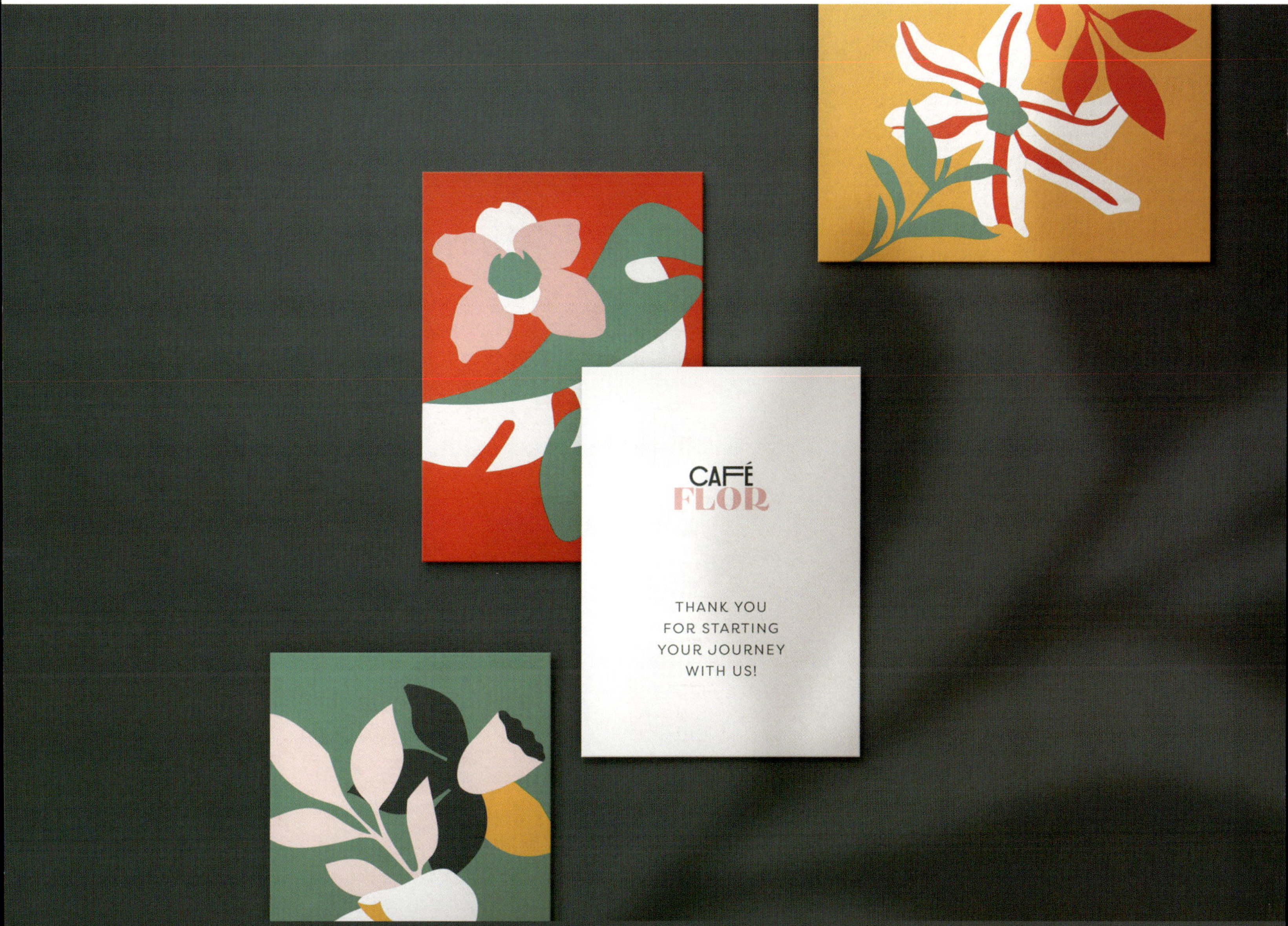
CAFÉ
FLOR
THANK YOU
FOR STARTING
YOUR JOURNEY
WITH US!

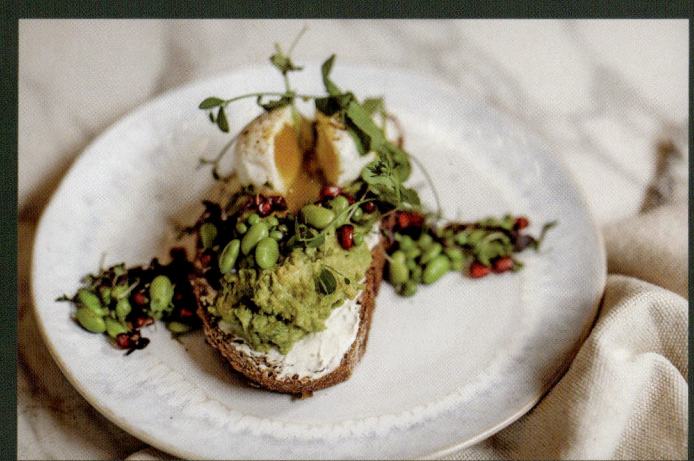

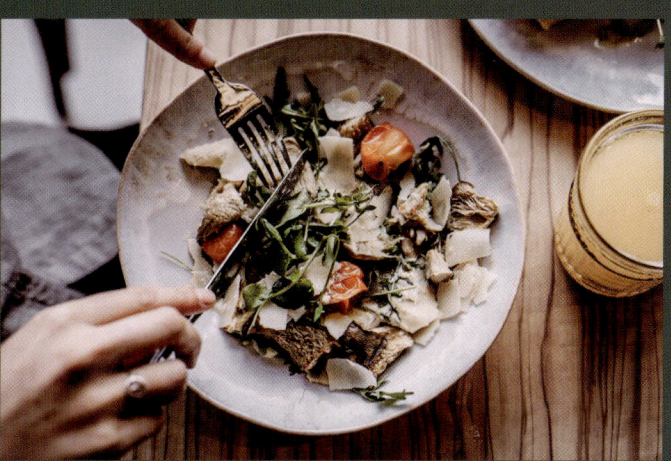

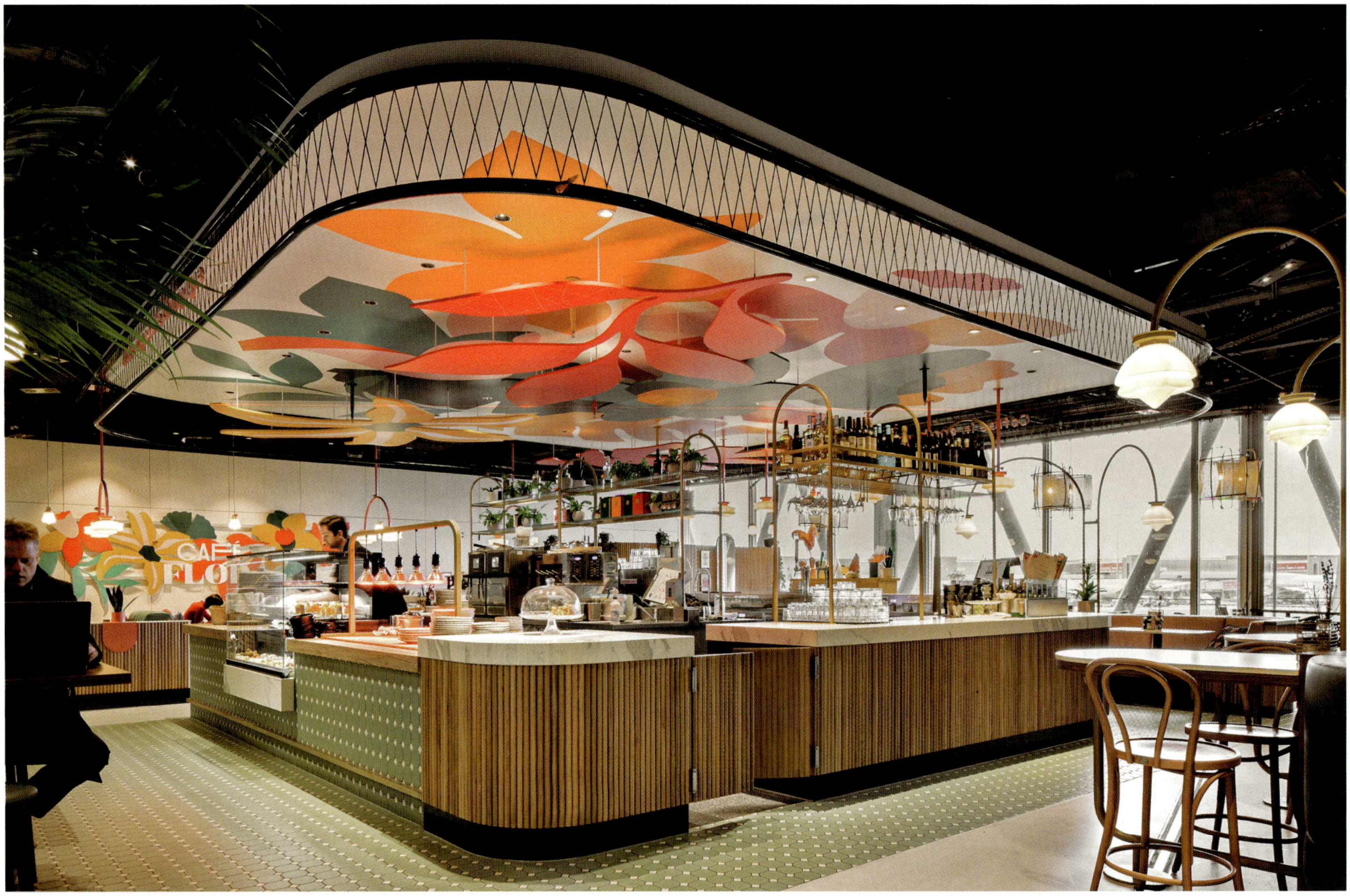
CAFÉ
FLOR

CAFÉ
FLOR
SANDWICHES
ALL DAY BREAKFAST
SALAD BOWLS
WRAPS
TREATS
DISHES
BITES
FOOD
DRINKS

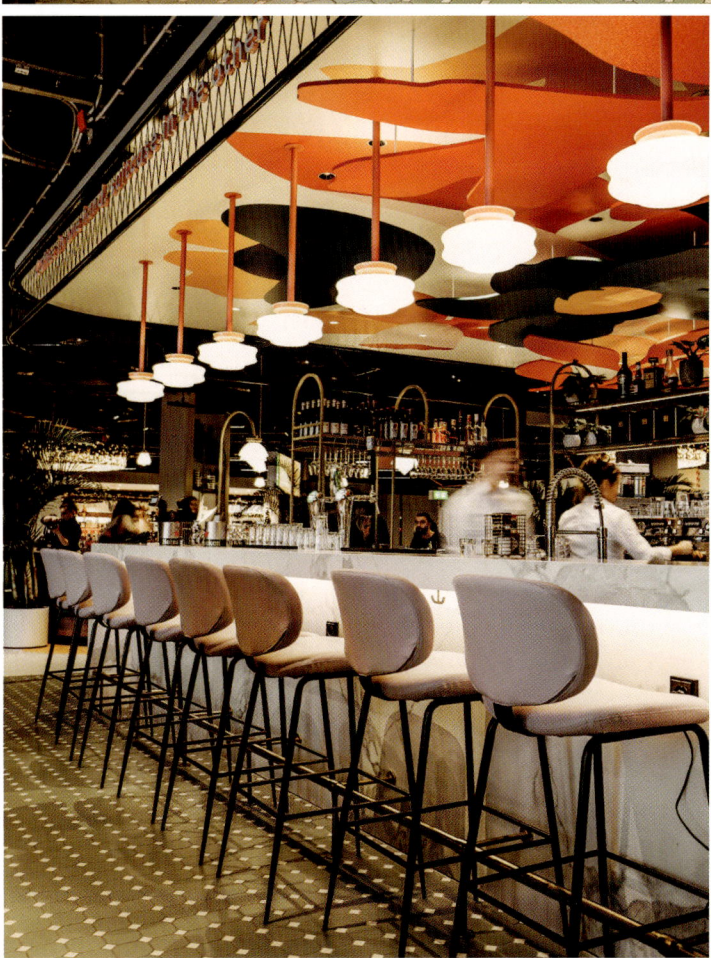

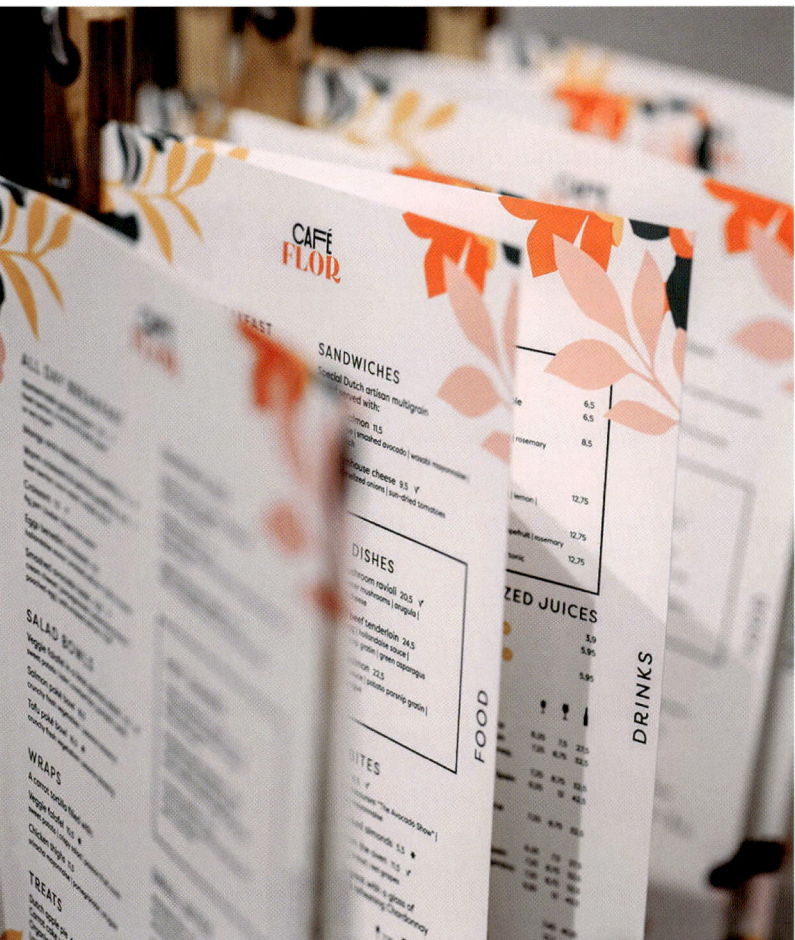

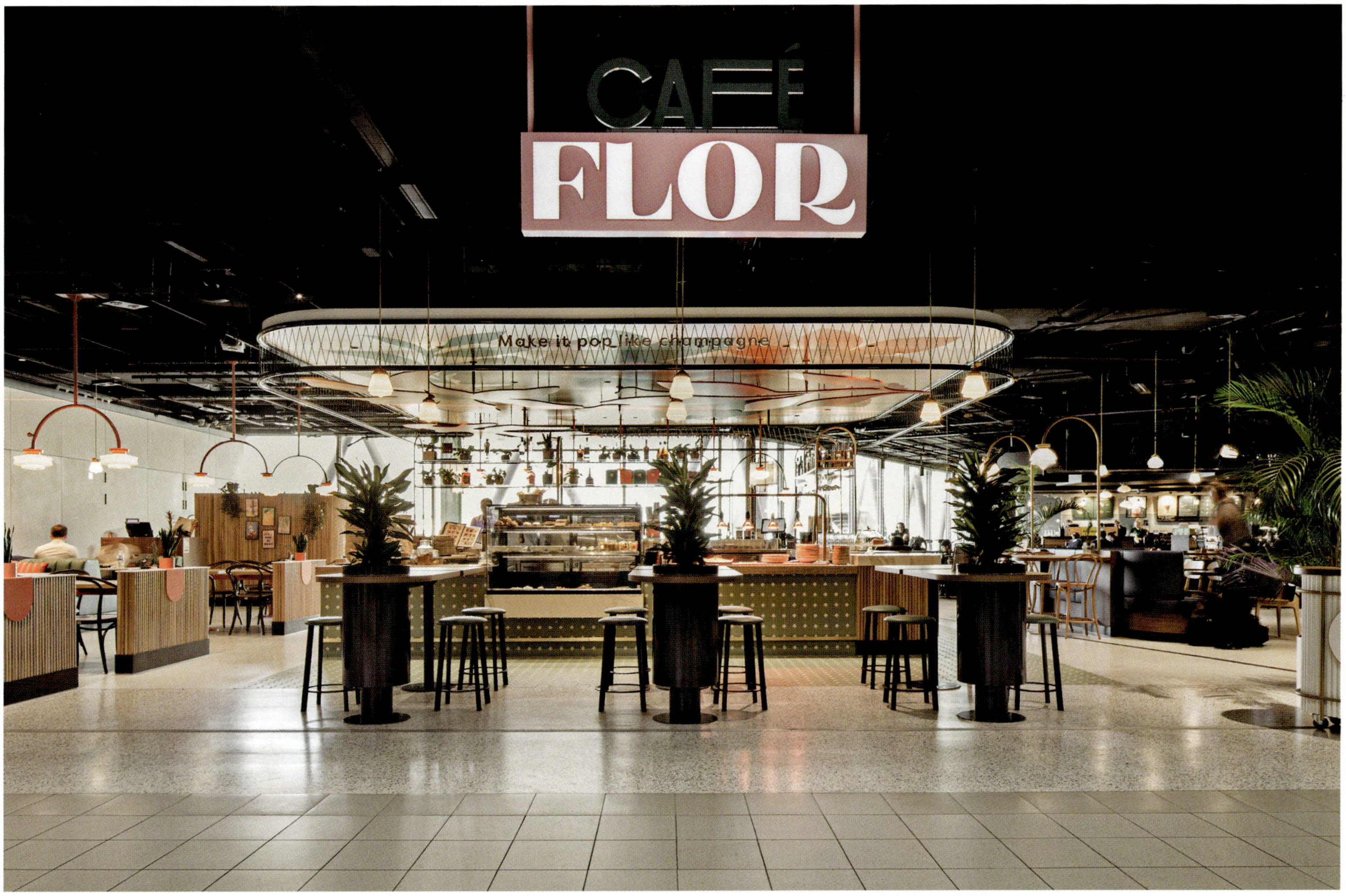
CAFÉ
FLOR
Make it pop like champagne

IT'S A KIND OF MAGIC
TOOKO

TOOKO

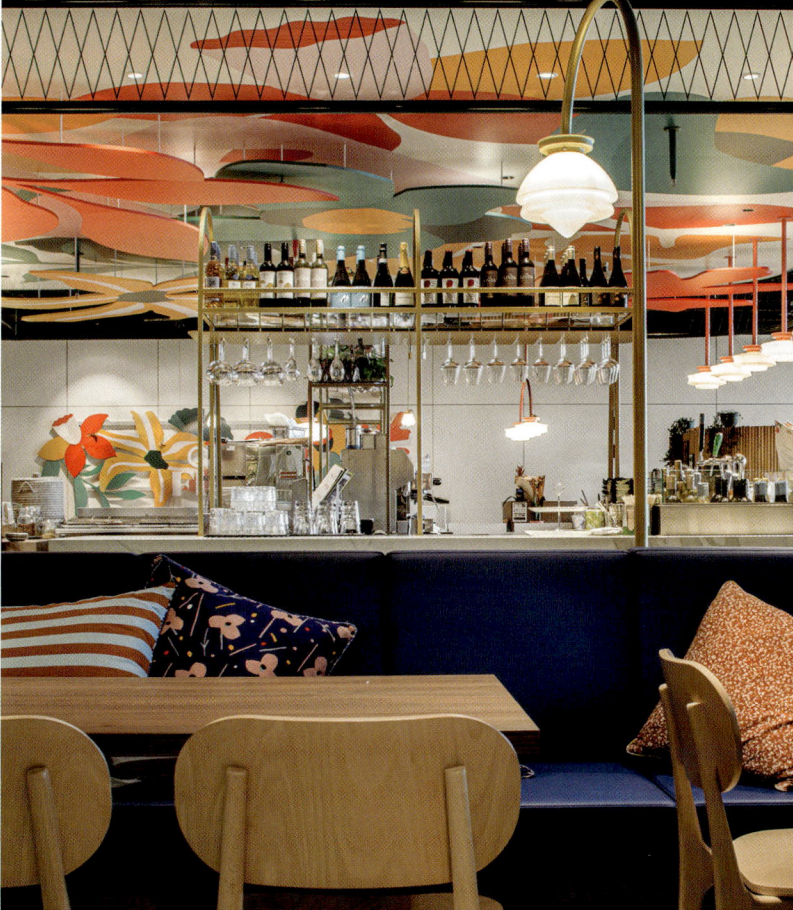

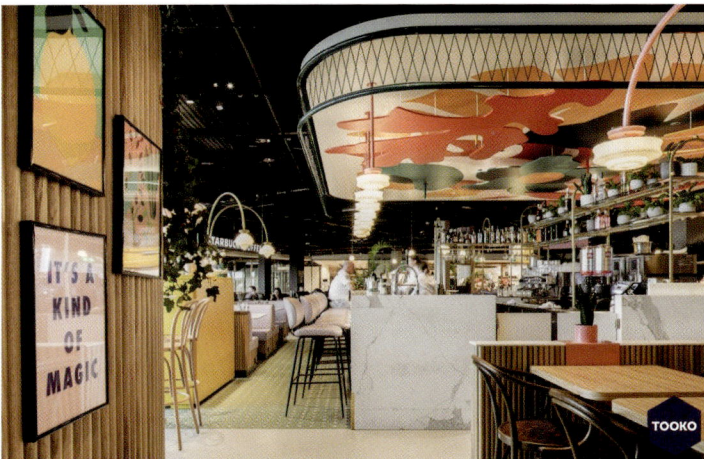

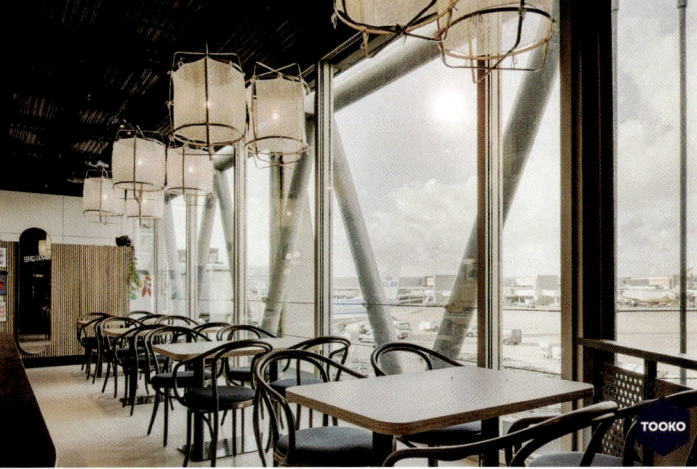

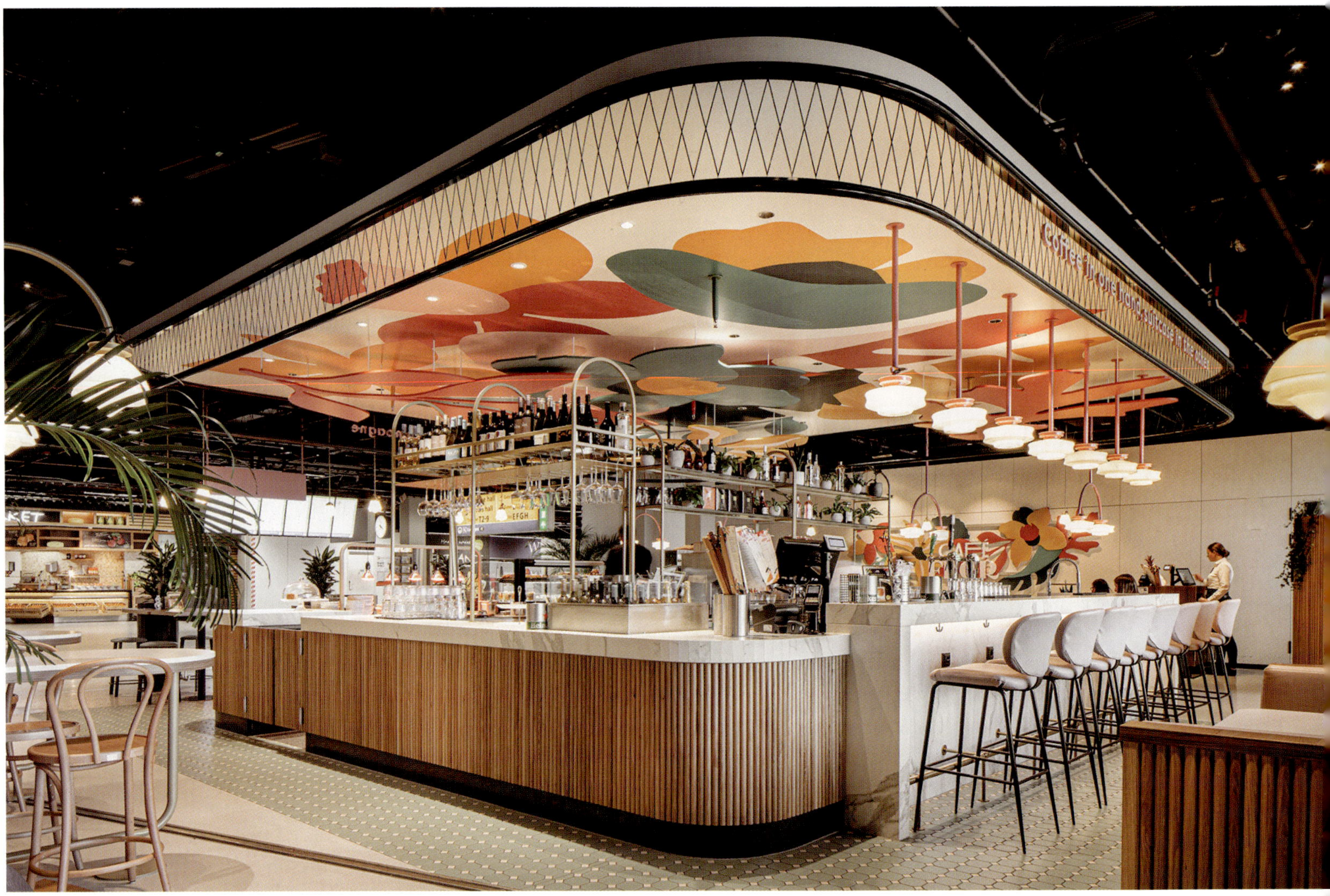

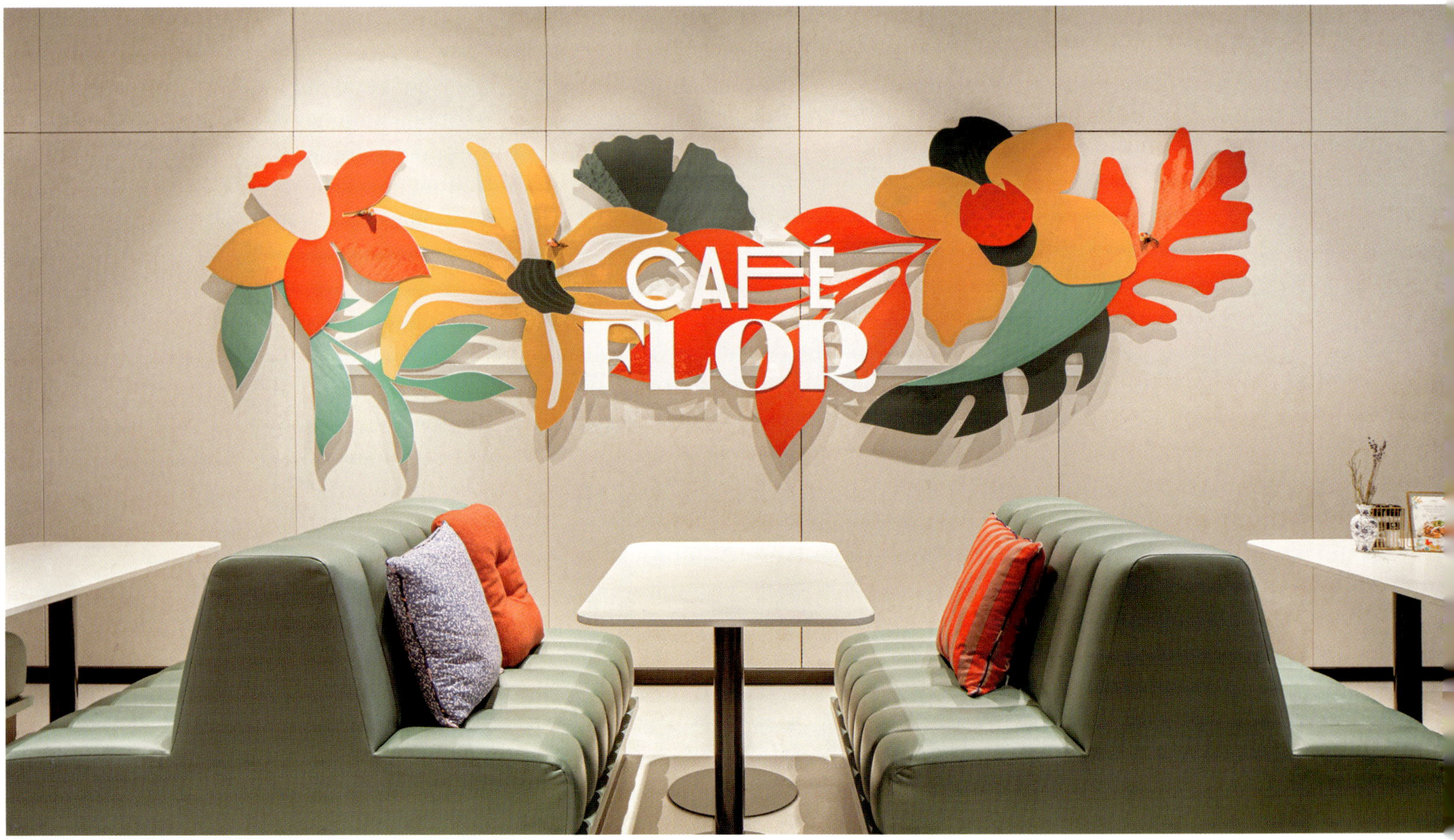

CAFÉ
FLOR

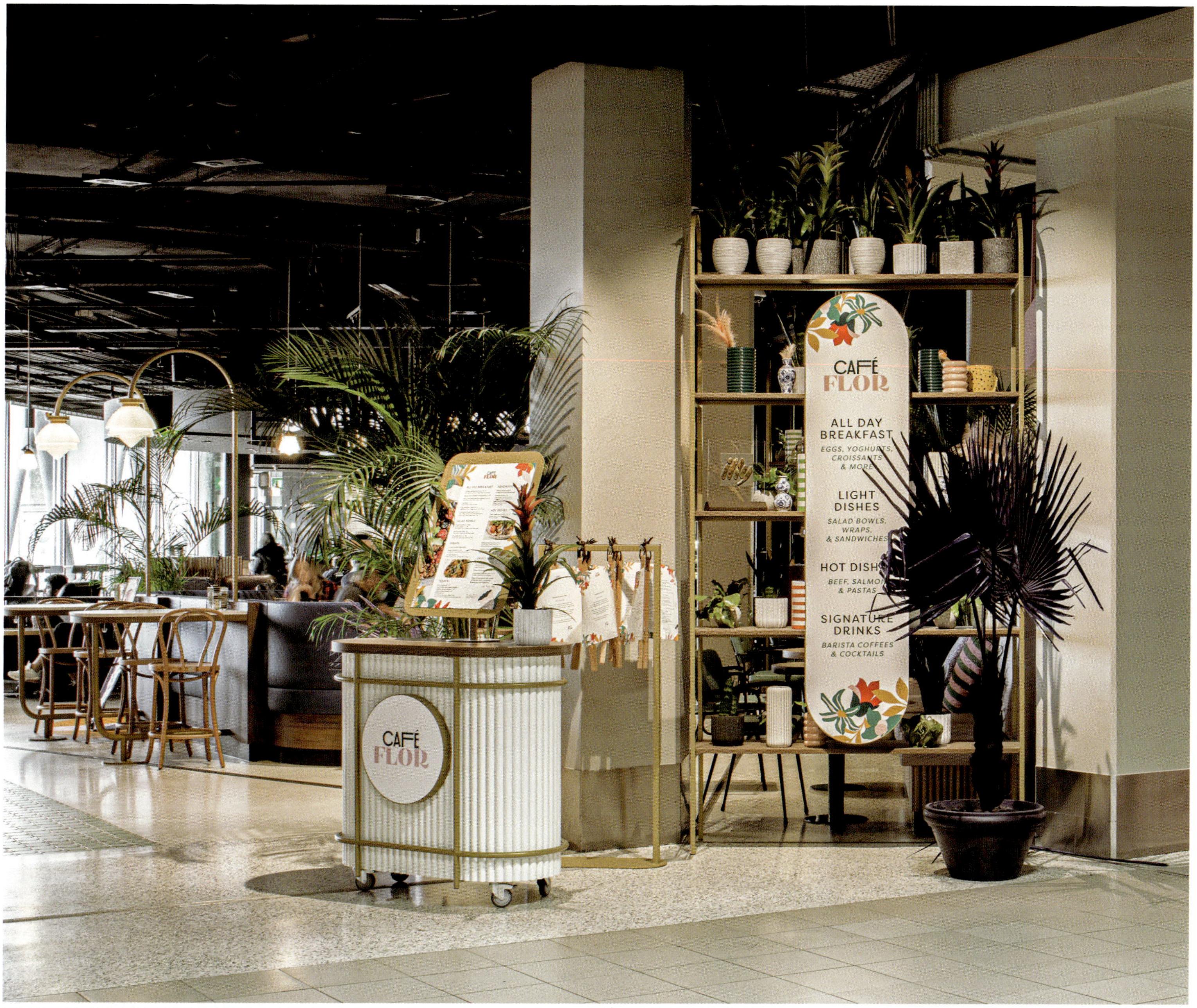
CAFÉ FLOR
ALL DAY BREAKFAST
EGGS, YOGHURTS, CROISSANTS & MORE
LIGHT DISHES
SALAD BOWLS, WRAPS, & SANDWICHES
SIGNATURE DRINKS
BARISTA COFFEES & COCKTAILS
CAFÉ FLOR

offee in one hand, suitcase in the other

HARVEST MARKET

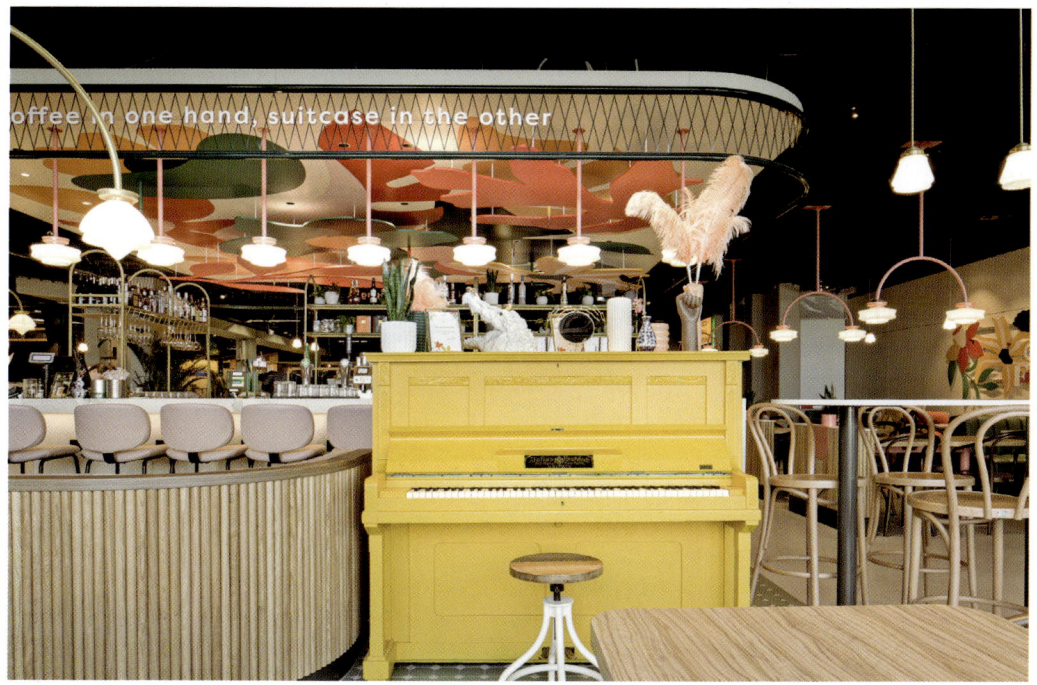

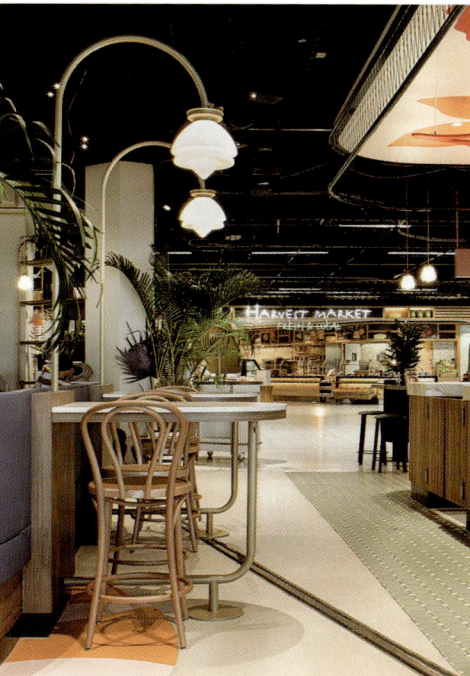

Barn's Cafe
Al Zahra Jeddah, Saudi Arabia

barn's
SINCE 1992

ODG designs?

The world leading Retail consultancy and communication group, established In 1983 and with offices in 30 countries. For 15 years, ODG team of Consultants, Planners, Creatives and Technical designers develop award winning retail programs and concepts in the world leading hubs for Operators, Retailers, Food service companies and brands. "ODG is a powerful mix of global expertise and professional consultants and designers. The team understands perfectly todays traveler needs and delivers creative solutions."

Design: *ODG designs*   **Homepage:** *www.odg-design.com*   **Location:** *Al Zahra Jeddah, Saudi Arabia*   **Image:** *BAYA Studio*
Chain store guide: *Al Tahlia Street in Jeddah, Jeddah International Airport, Jeddah Formula 1.*

Since before the pandemic, ODG has been collaborating with Barn's to design multiple outlets. ODG designed Tahlia Street branch in Jeddah along with other flagship and exclusive branches such as their new spot at the Jeddah International Airport – Arrivals, the Al Zahra branch and the exclusive location at the Formula 1 racetrack. It's well known that Saudi Arabia has a robust coffee culture in the GCC if not in the entire Middle East. From world class baristas to lush spaces and exclusive specialty blends and beans, Saudi is the place to discover new coffee tastes in design award worthy spaces. Because of that, many home-grown coffee shop chains have established in the kingdom, across multiple formats, from stand-alone coffee shops to curb-side drive through units, all united by the love of coffee.

Barn's Cafe is the first contemporary yet fully homegrown coffee concept in KSA, founded in 1992. With its distinctive roasting facilities, they guarantee quality standards across all coffee products. Barn's café remains synonymous with a unique image of excellence and product quality in the mind of their loyal customers. The design approach put in the center not just the exceptional quality of the coffee but also to bring the community together, to learn more about the authenticity and the history of Barn's over the years. The space is bright and friendly, a place to visit every day, to have a chat, work or just to spend a good time while drinking coffee.

Jeddah International Airport

148

since 1992

Jeddah Formula 1.

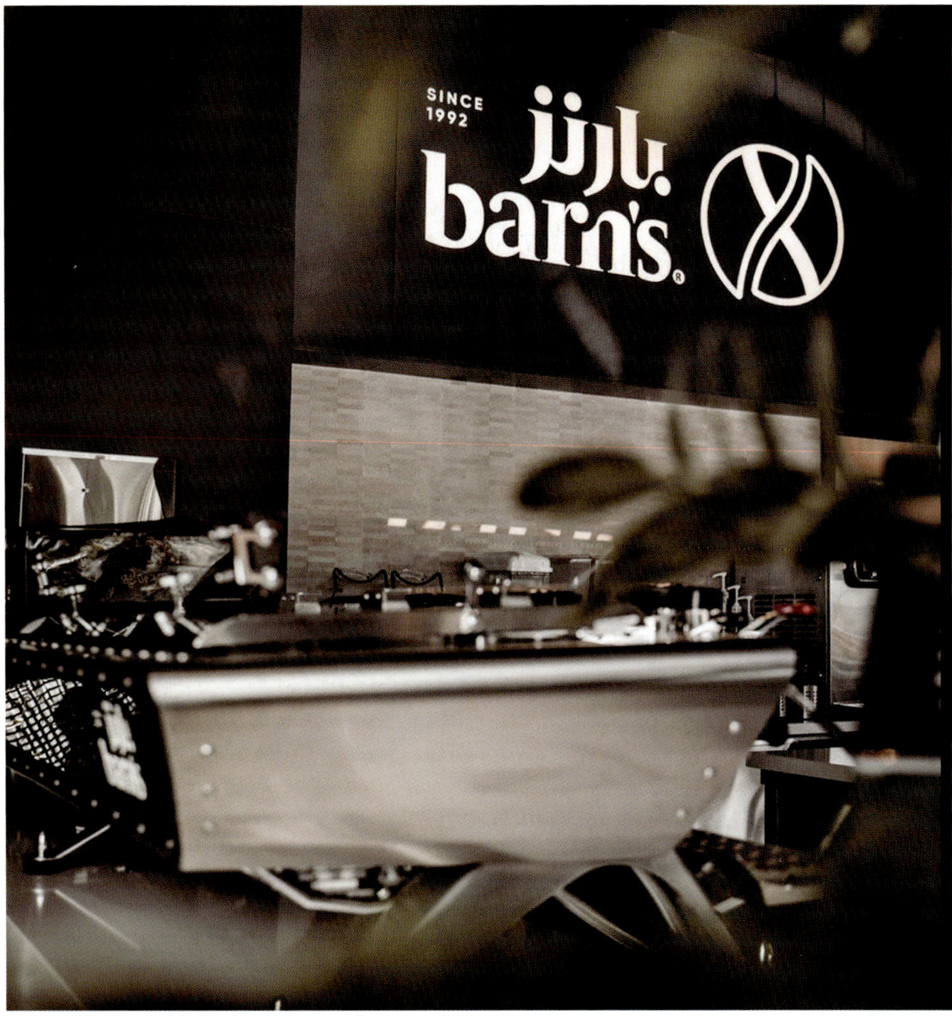
SINCE
1992
بارنز
barn's

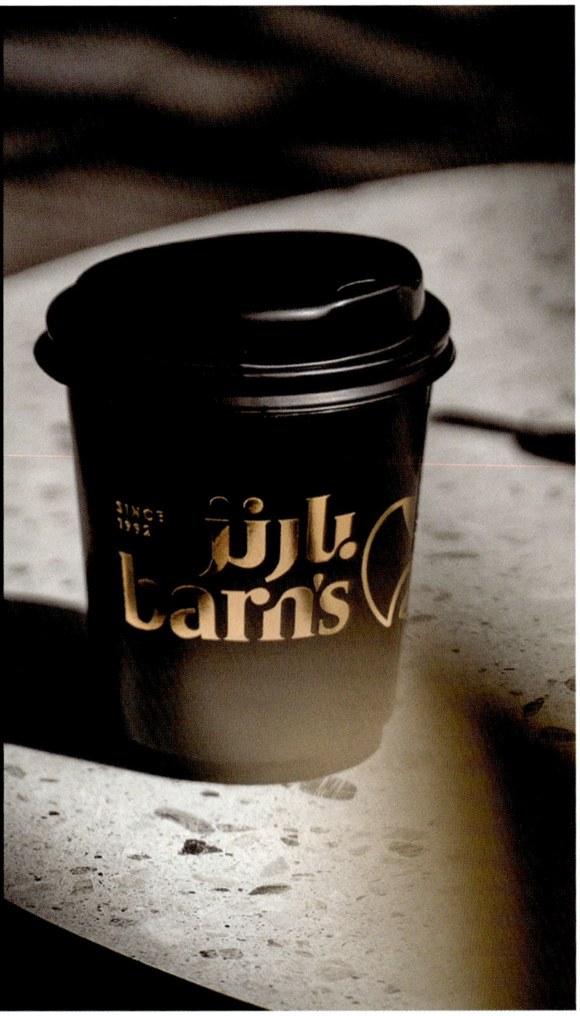
SINCE
1992
بارنز
barn's

بارنز
barn's
بارنز
barn's
بارنز
barn's

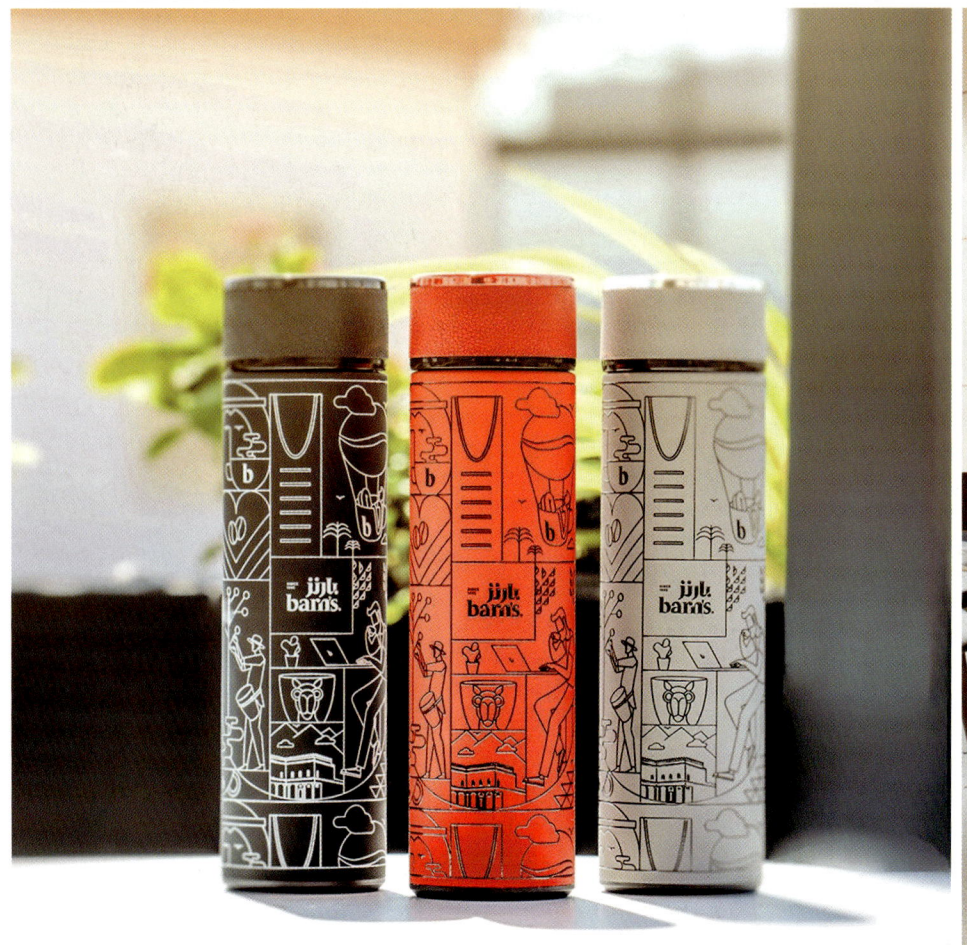

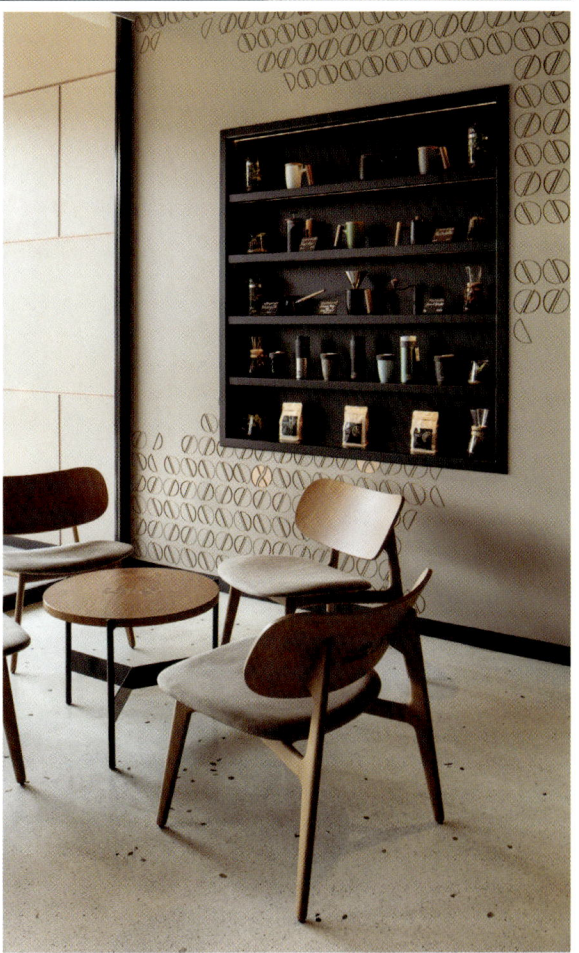

Al Tahlia Street in Jeddah

155

La Ganea
Villagana BS, Italy

Studio Mabb?

Studio Mabb è un gruppo di architetti altamente specializzato in Interior la cui ricerca progettuale ha come cardine la valorizzazione dello spazio attraverso forti caratteri materici e puntuale studio della luce. Fondamentale risulta la continua collaborazione con laboratori e atelier artigianali al fine di creare un' ambientazione unica e totalmente su misura, come un prodotto di sartoria d' interni. Ogni progetto è il risultato di un calibrato studio di materiali e finiture, luci e ombre, funzioni e sensazioni, il cui scopo è quello di rendere un ambiente un luogo emozionale, fisico-tattile, in cui il corpo diventa elemento sensibile.

Design: *Studio Mabb* **Homepage:** *www.studiomabb.it* **Location:** *Villagana BS, Italy* **Area:** *200 ㎡*
Team: *Andrea Baselli, Riccardo Belinci, Sara Magnone....* **Photographer:** *Carola Merello, Pietro Dardano*

The 16th century Palazzo Martinengo is today hosting Ganea: a place devoted to beer, tasty food and pizza. Ganea is the ancient name for Villagana, which in Latin meant tavern, inn, point of rest for travelers and merchants. It has been placed in the side wing of the property, which in the past was dedicated to stables.

The estate is subjected to specific restrictions by the local artistic and cultural heritage superintendence, hence the layout of the venue is the original one. Since the internal walls are also under the same restrictions, wiring and lighting systems are attached directly on the walls, fixed with brownish iron bars made by local artisans and inspired by master Sigurd Lewerentz.

The materials selected for the project are mainly iron, cotto tiles and conifer wood as an historical reminiscence of the memory of the estate and its surrounding region. The iron frame concept recurs in every room of the project: from the big framed chandelier in reinforced glass, to the restaurant booth located in the last room conveniently hiding restrooms entrances, as well as the inner door in the main entrance. In the first room a bottle rack made of iron and reinforced glass creates a scenographic backdrop to the natural iron monolithic counter, as well as conveying a sense of perspective and direction to the room. In the main room the big glass windows with iron frameworks overlooking the garden are authentic, faithfully and genuinely restored to their original design, while the openings of the other rooms are made in larch wood.

In the two small cozier rooms there are two long benches and tables made in brushed fir and big festive tables which are instead imported from Asia, with the purpose of highlighting the historical link between Villagana and the Asian continent. In fact, during ancient times Villagana had the privilege of owning a pier on the Oglio river bank, which used to connect the Venetian Republic to the Duchy of Milan.

Lots of goods, mainly silk, used to be shipped by such Pier as documented by the numerous original evidences written in Asian languages, originally kept in the library and nowadays exhibited in the heart of the venue. The unique handmade baked tile flooring illustrates the consecutive time eras of the venue thanks to its several chromatic shades

The sandblasted ceiling beams and the rustic lime-based plaster of the walls give to the space a peciliar earthy component. Some ancient maps of Martinengo Counts domain are hanging on the walls next to the illustrations of Czech artist Vlastimil Košvanec. A museum situated on the main floor exhibits more of his art. The Ganea is a venue in which the project gives life to the history of the place.

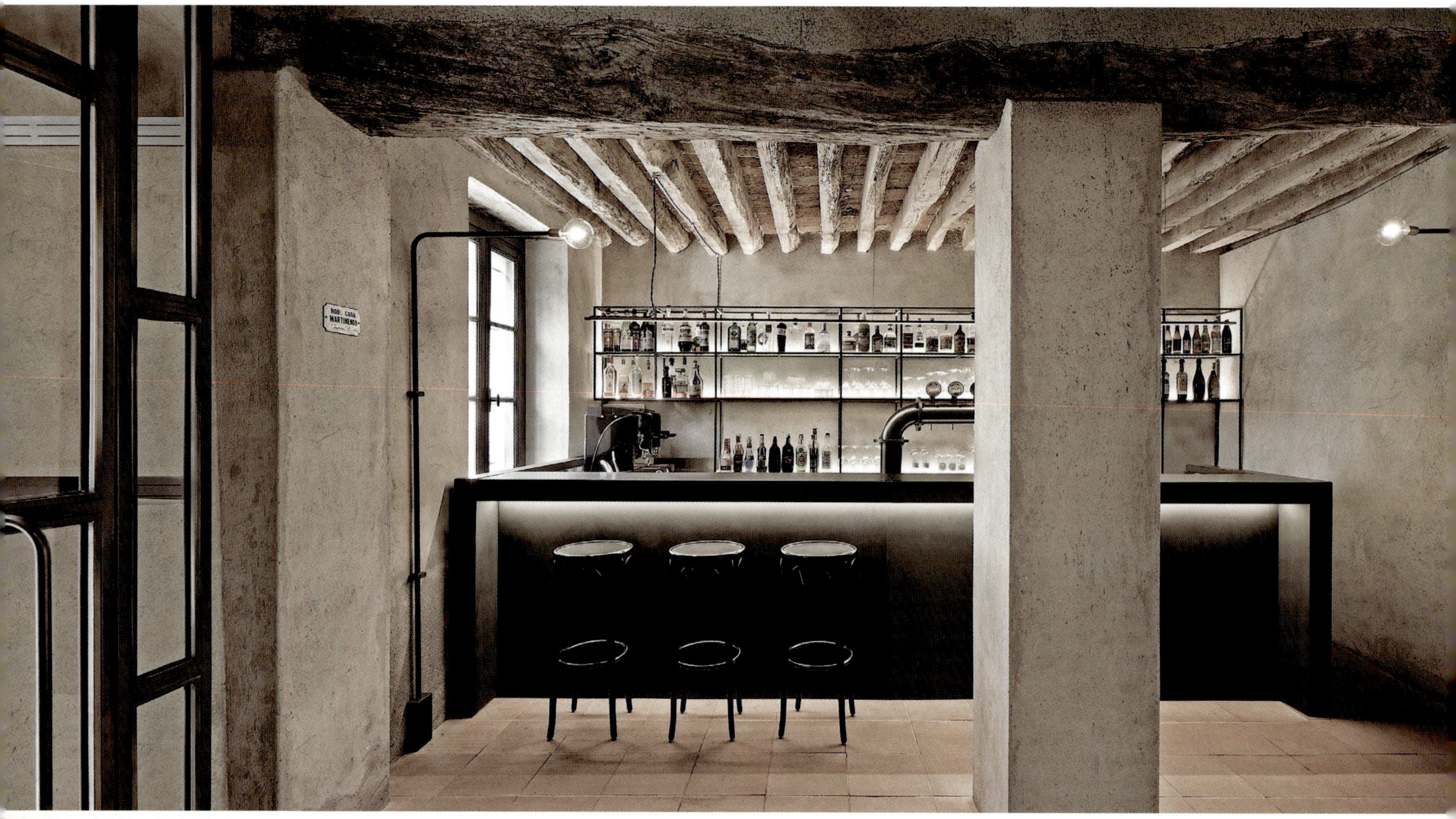

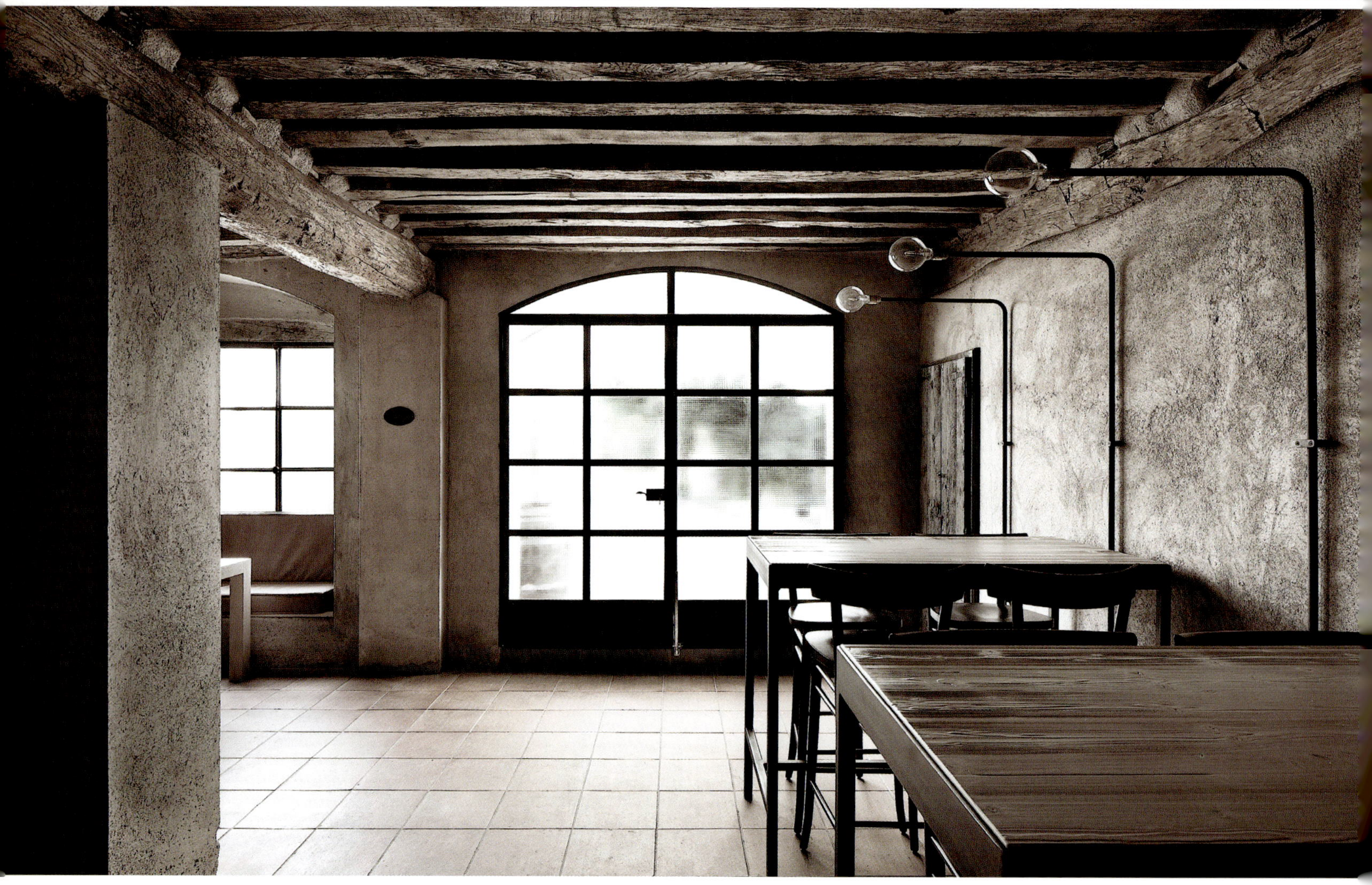

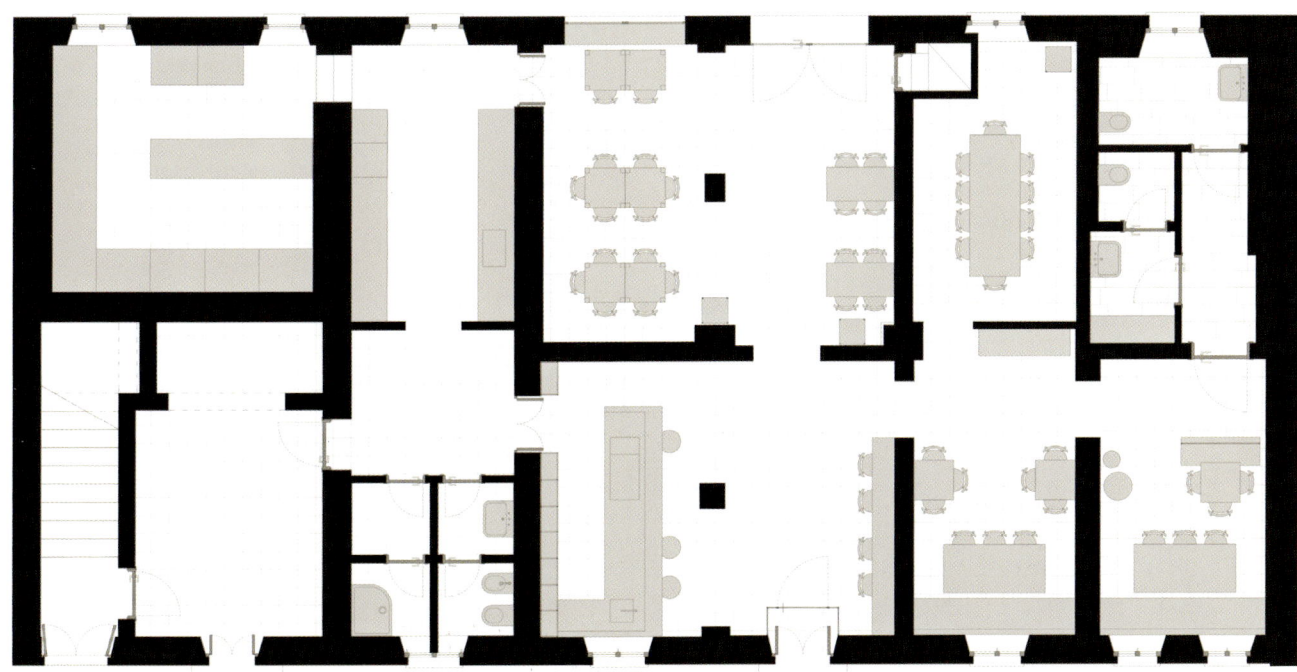

Floor Plan

159

Space+Craft (Interior design)?

Breathtaking design from forward-thinking architecture and interior design professionals. With eight years of combined experience in their fields, the co-founders of space+craft take pride in offering premium services to a broad range of discerning clients.

Function design consultant: Pittawas Euavongkul **Graphic Design:** Katekamon Thummasaen **Location:** Bangkok, Thailand
Interior design: space+craft **Photographer:** Ketsiree Wongwan **Area:** 100 sq.m

Kaizen coffee co. is known as one of the best specialty coffee places in Bangkok, founded by a young barista, Arnun Wattanaporn whose passion is highly influenced by Australia's vibrant coffee scene. After a widely successful reception from its coffee lovers, Kaizen has recently launched its new home in Ekamai, one of the prime locations in the heart of Bangkok.

Surrounded by multiple high-rise buildings in some of the most vibrant areas in Bangkok, the double space stand-alone establishment of 100 sq.m. was raised among them. Having the front façade decorated with high clear glass, the first architectural challenge was to make the interior design outstanding and attractive when viewed from the main street. An enormous round oak plate engraved with 'K' for Kaizen, supplemented with LED light, was then affixed on the back wall accordingly. This element can be easily observed from the main street and thus presents a very firm identity to the café. Unlike the previously owned Kaizen branches, the café owner aims to enhance further essence of coziness rather than a typically modern and minimal café atmosphere.

This new Kaizen will not serve only a variety of coffee and beverages, but also a wide range of scrumptious healthy meals freshly prepared and served daily during breakfast till dinner. This intention has changed Kaizen's interior design concept in both style and functions. The natural finish oak woods and dark grey bricks wetre opted as the main principle materials. The result is the harmonious combination and beautiful contrast of the structure at once. While the oak woods transmit a sense of coziness, the dark grey bricks bring about the sophistication to the interior space. The distinct brick pattern was installed at the counter bar and the first floor wall, adding the design details as well as the exceptional character to the space.

The furniture layout was designed as an open plan with a functional twist on both first and second floors. This was done to propose a variety of seating groups for the clients to have options of their preferred table depending on their desire whether they will have only coffee or meals. Finally, the new look of Kaizen Coffee has become uniquely outstanding among various coffee places in town and completely changed the well accepted image of Kaizen, which is one of the best providers of quality coffee as well as healthy meals in the capital city of Thailand.

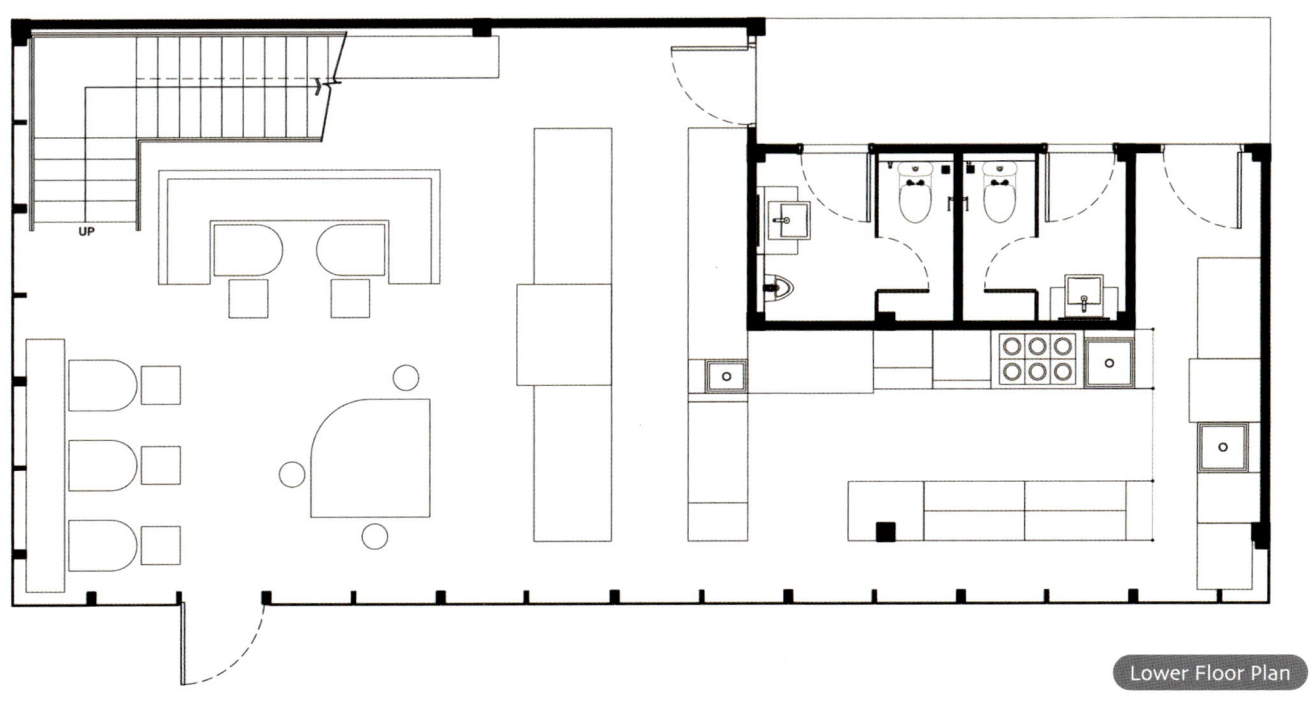

Lower Floor Plan

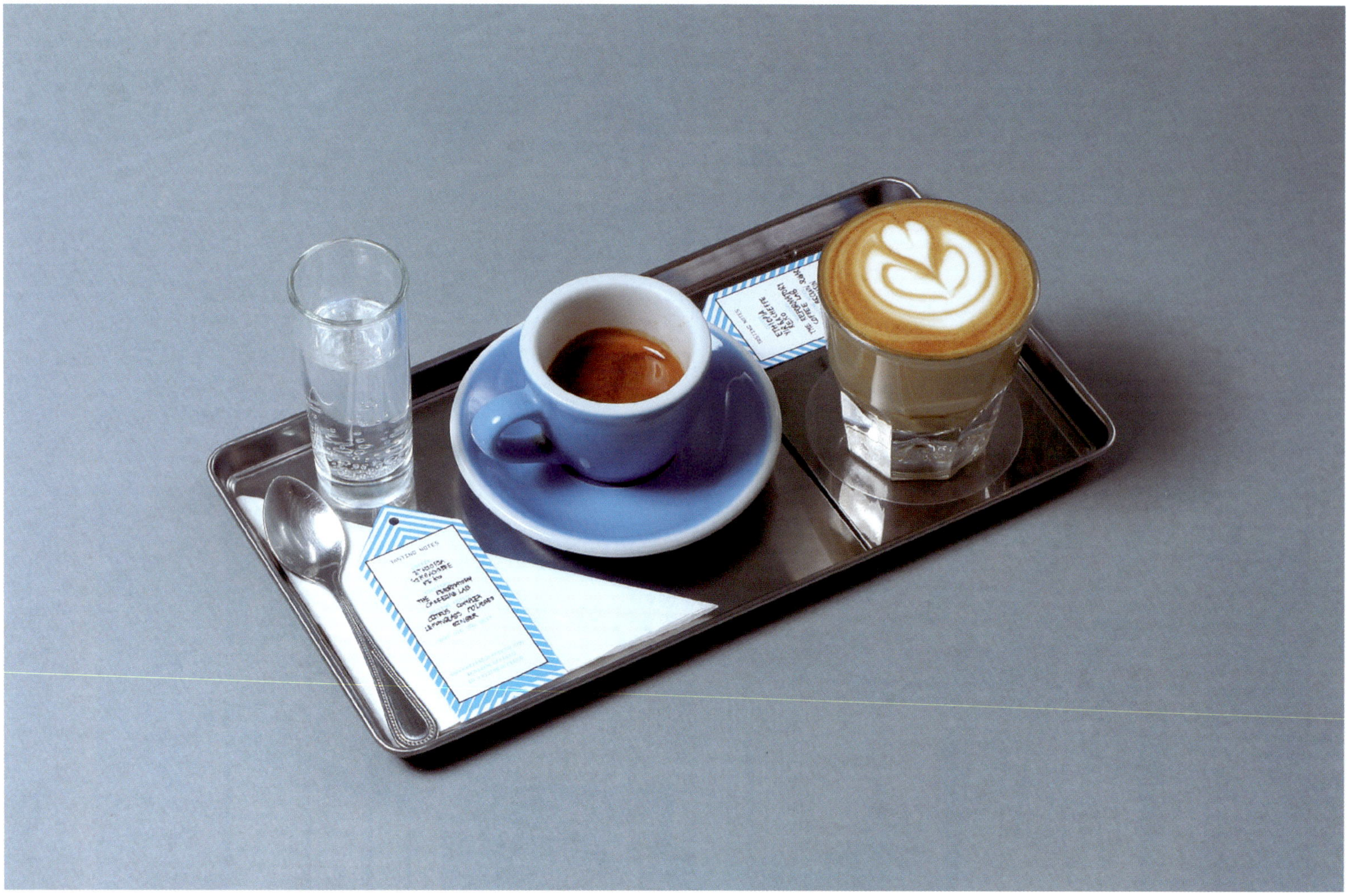

KAIZEN
COFFEE

NEW HORIZONS

KAIZEN
COFFEE CO.
ARNUN WATTANAPORN
FOUNDER/MANAGING DIRECTOR
KAIZEN COFFEE CO.

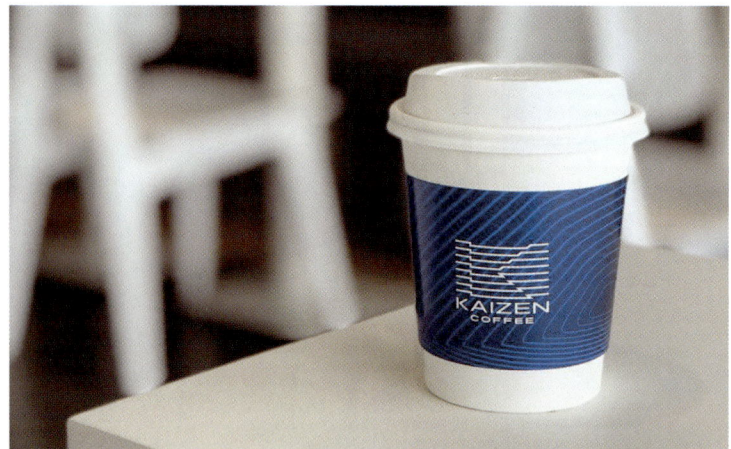

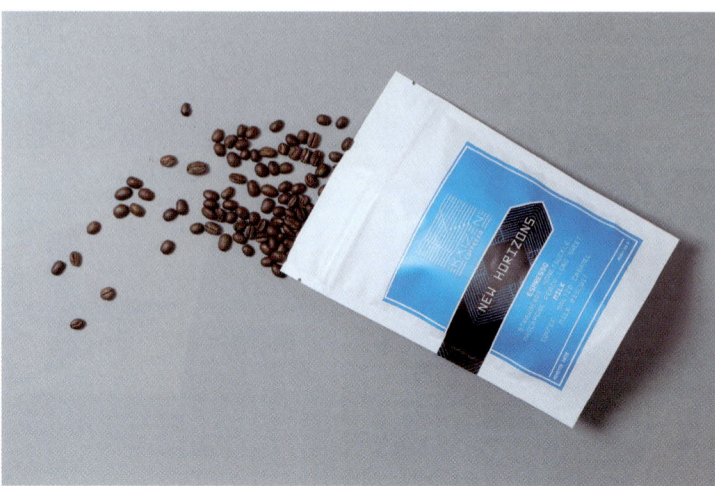

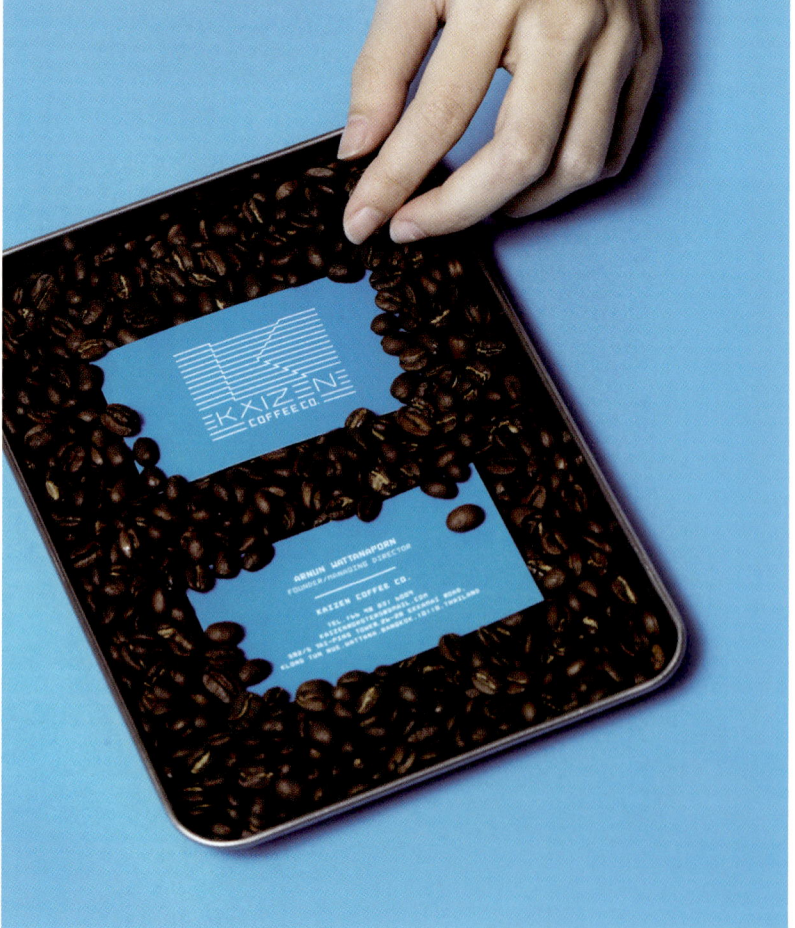

PROGRESS OVER PERFECTION
NEW HORIZONS

NEW HORIZONS
ESPRESSO
STRAWBERRY HONEYSUCKLE
MASCAPONE PEACH LONG SWEET
MILK
TOFFEE SALTED CARAMEL
MILK BISCUIT

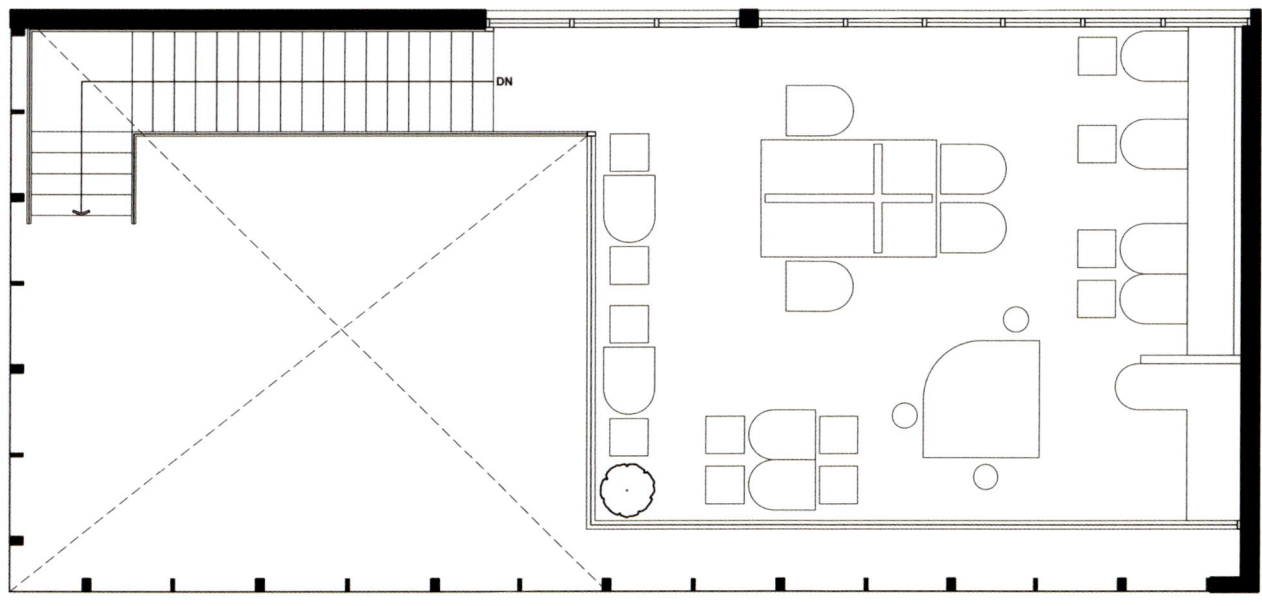

Upper Floor Plan

KAIZEN
FILTER ROASTED COFFEE
GUATEMALA
El Injerto
RUBY MALAWI GEISHA

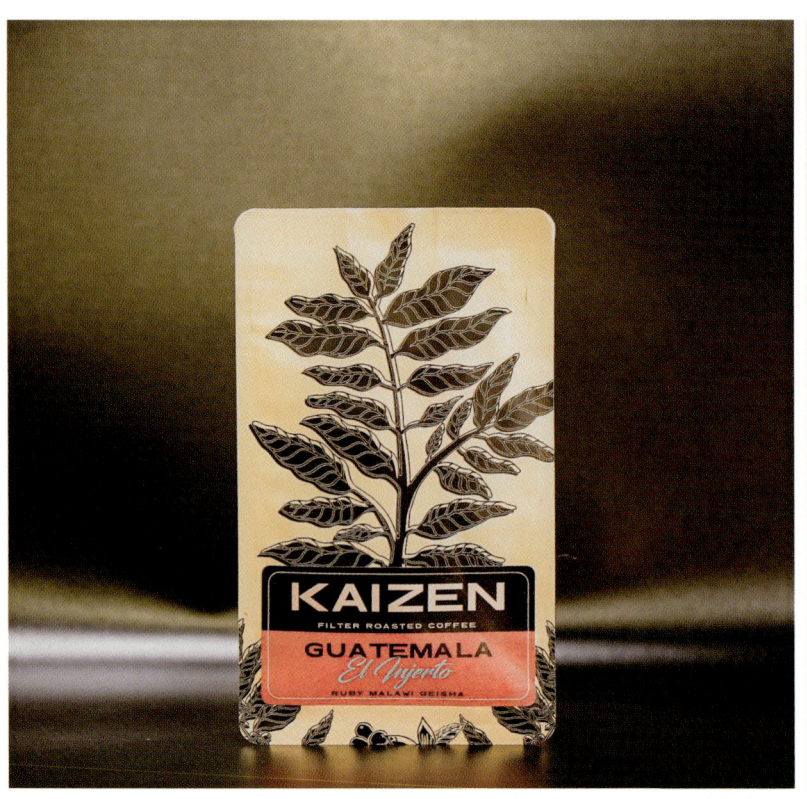

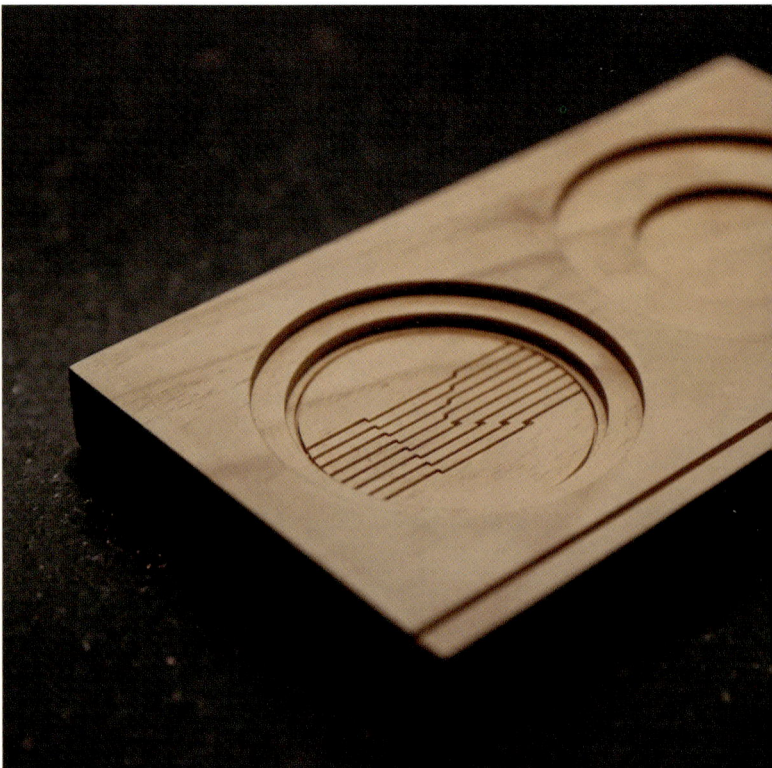

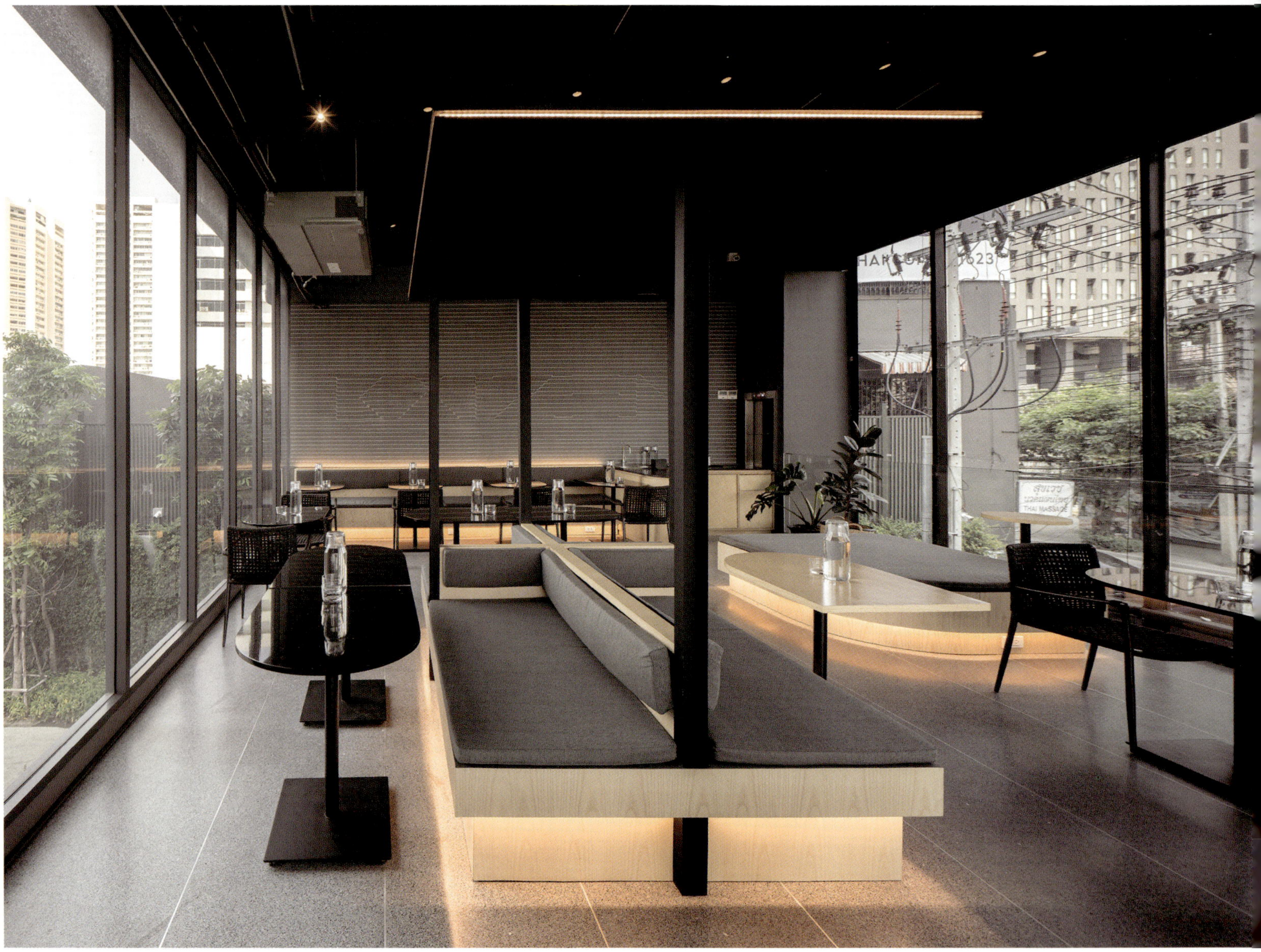

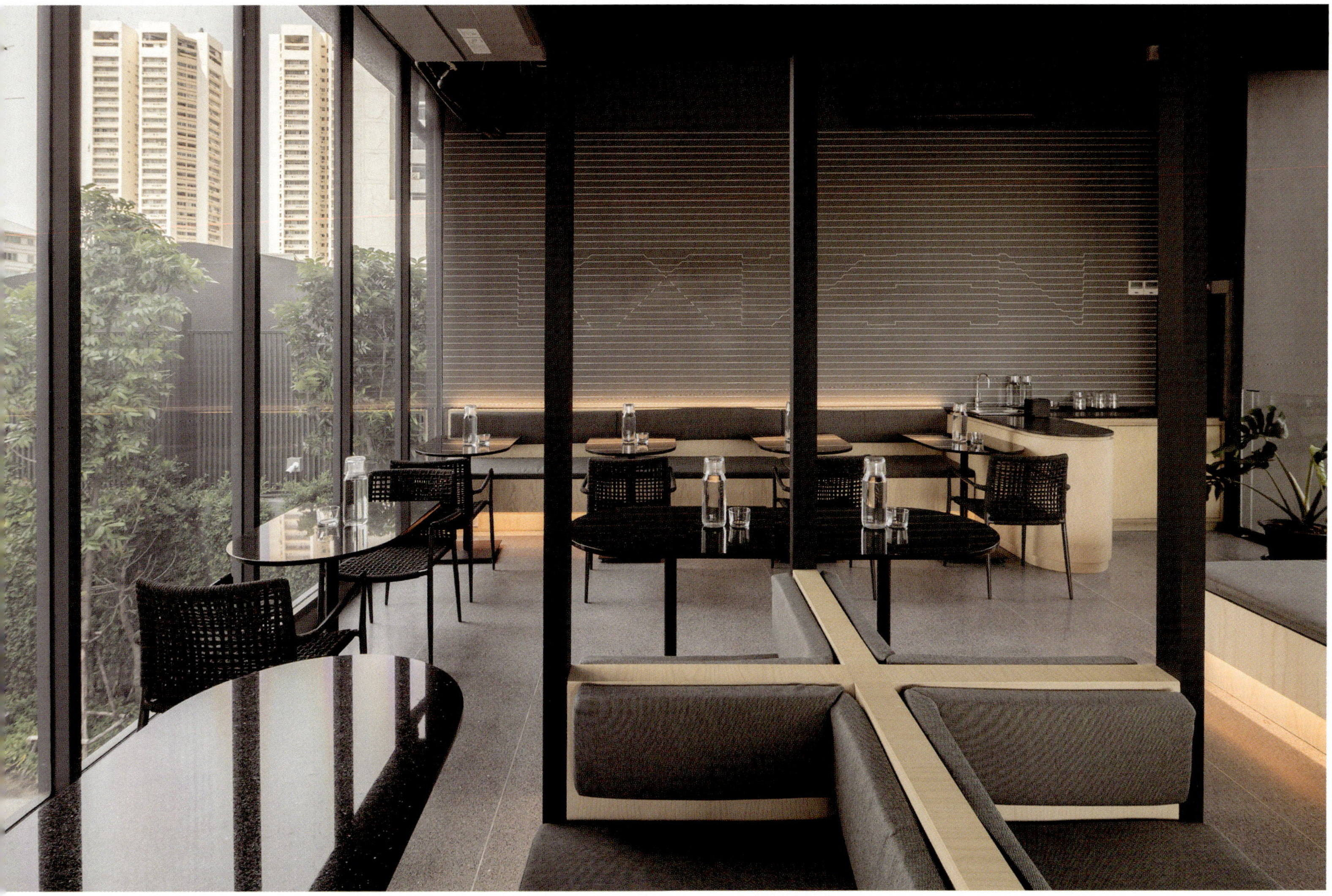

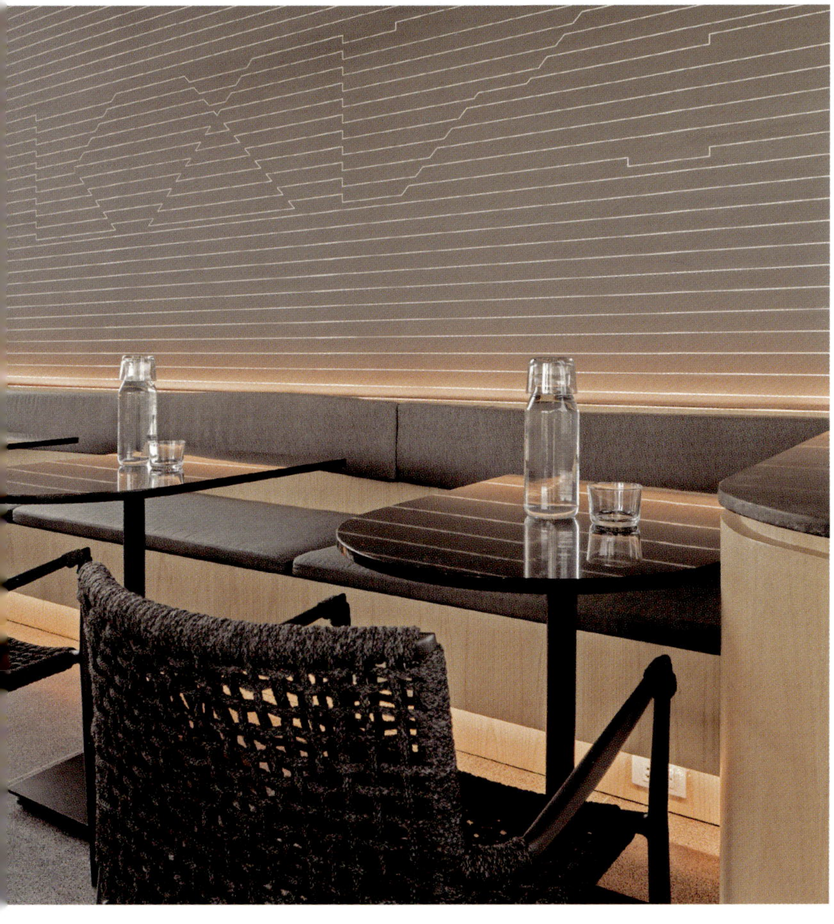

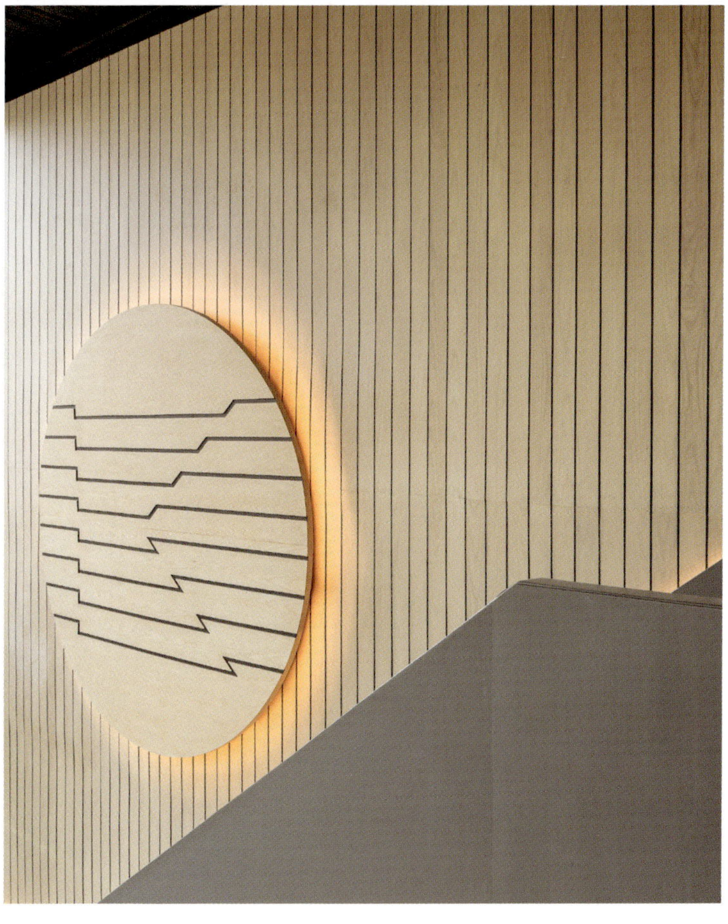

GAGA King Glory Cafe
King Glory Mall - Shenzhen, China

gaga 鲜语®

MA design?

We are an Architecture, interior design & space planning partnership group, in EGYPT & BAHRAIN passionate about creativity and cutting edge corporate design. MA design w.l.l. (interior / Graphic) is A Bahraini Company founded & owned by Mohammed Amer, Egyptian Interior Architect page browse some of his art works & projects, plus a range of magnificent interior design.

Design: *MA design* **Homepage:** *http://www.madesign1.com* **Location:** *Shenzhen, China* **Area:** *525 sq.m*
Team: *Vega Li, Yu Yin, Ryan Yang* **Photographer:** *Coordinayion ASIA*

An oasis of tranquility inside one of Shenzhen's most bustling shopping malls, this newest addition to the Gaga family of restaurants and cafés combines inspiration from the natural world with a focus on flexibility. Striking a balance between inside and out, momentum and calm, Coordination Asia's eye-catching interiors create an eatery for all hours, and every kind of customer.

Coordination Asia's ongoing collaboration with Gaga restaurants and cafés continues, with a new venue inside King Glory Mall, Shenzhen. With dramatic curved lamella ceilings, natural materials throughout, and a color palette of midnight blue, teal, and aquamarine, it evokes the tranquility of an idyllic beach – in the heart of one of China's most buzzing metropolises!
Imagined as both an oasis for shoppers, as well as an inspiring setting for meeting with friends, catching up on emails, or simply relaxing, Gaga King Glory's expansive entrance and floor-to-ceiling glass façade draws visitors into the space. With attractive outdoor seating for some 64 diners, as well as tables extending into the mall itself, the design blurs the boundaries between inside and out, establishing the venue's relaxed, informal atmosphere from the outset.

Connecting both Gaga's interior and exterior elements is a lamella ceiling in bright white aluminium. Reminiscent of a crashing wave, it signals a subtle beach theme that runs throughout the design, while its dramatic curves hint at the flexibility of the wider venue. Inside, the 400sqm interior appears as an ocean of blues, turquoises and greens. Designed with versatility in mind, and to match a full spectrum of café-goers' preferences and needs, Gaga inside King Glory Mall incorporates several distinct spots, accommodating a total of 150 guests. Each with their own unique ambience and feel, spaces are distinguished through furniture styles, lighting types, and most striking of all, alternate terrazzo and parquet flooring,
They include a central raised lounge area: flanked by low-level tables, plush banquette seating and contemporary chairs, it also features high tables and stools. Moving closer towards the cornflower-blue counter, eye-catching cylindrical benches offer a more informal spot for those grabbing a quick coffee on the go. For cozy catch ups, two additional lounge areas see comfy sofas and statement lighting create a home-from-home for respite and relaxation; while long communal tables are ideal for groups, be they drinking, eating, or working. Throughout, pendent lighting in burnished bronze and black lend an intimacy to the space, while adjustable LED lights concealed inside the lamella ceiling carry the eatery through day, to dusk, and night.

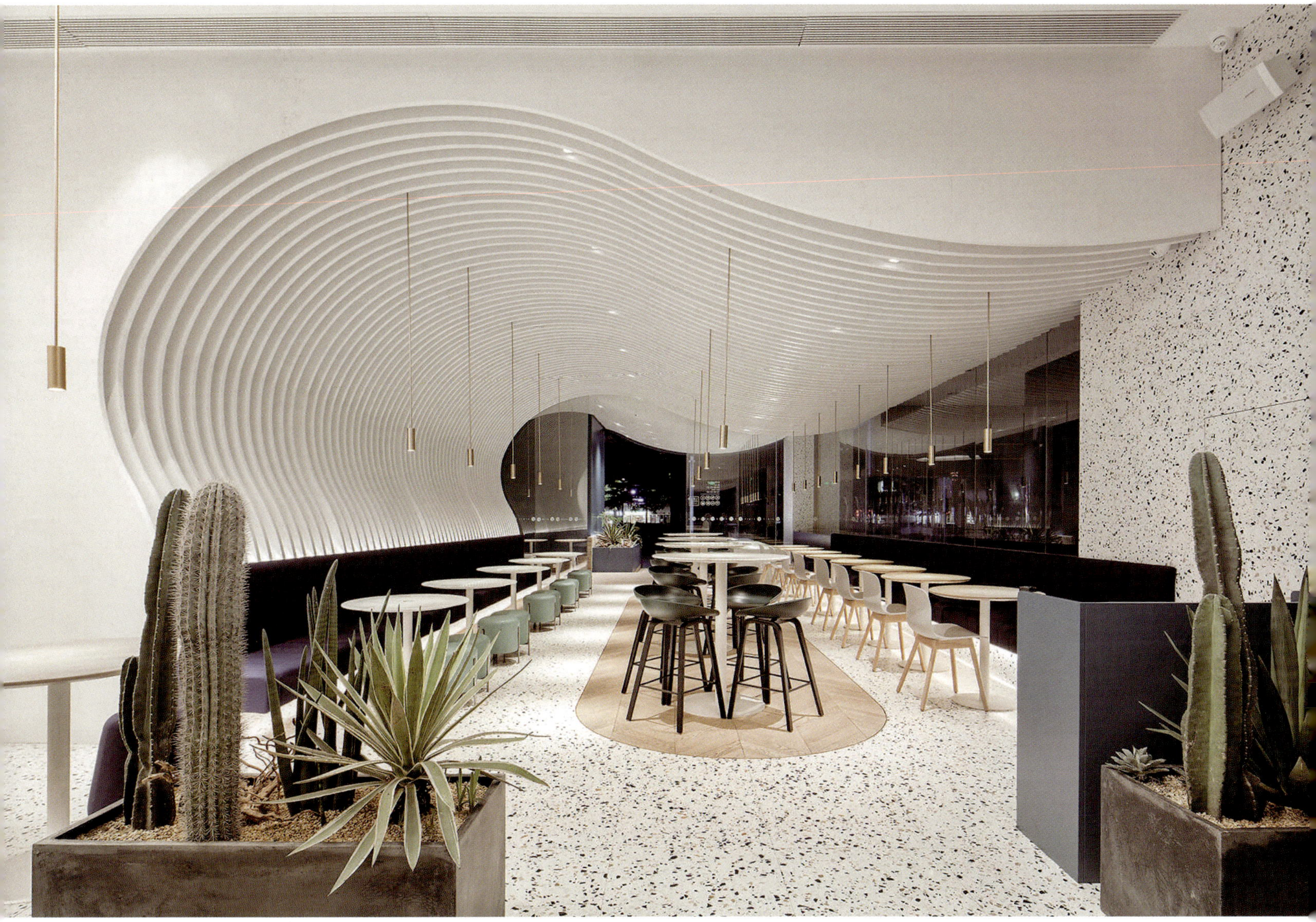

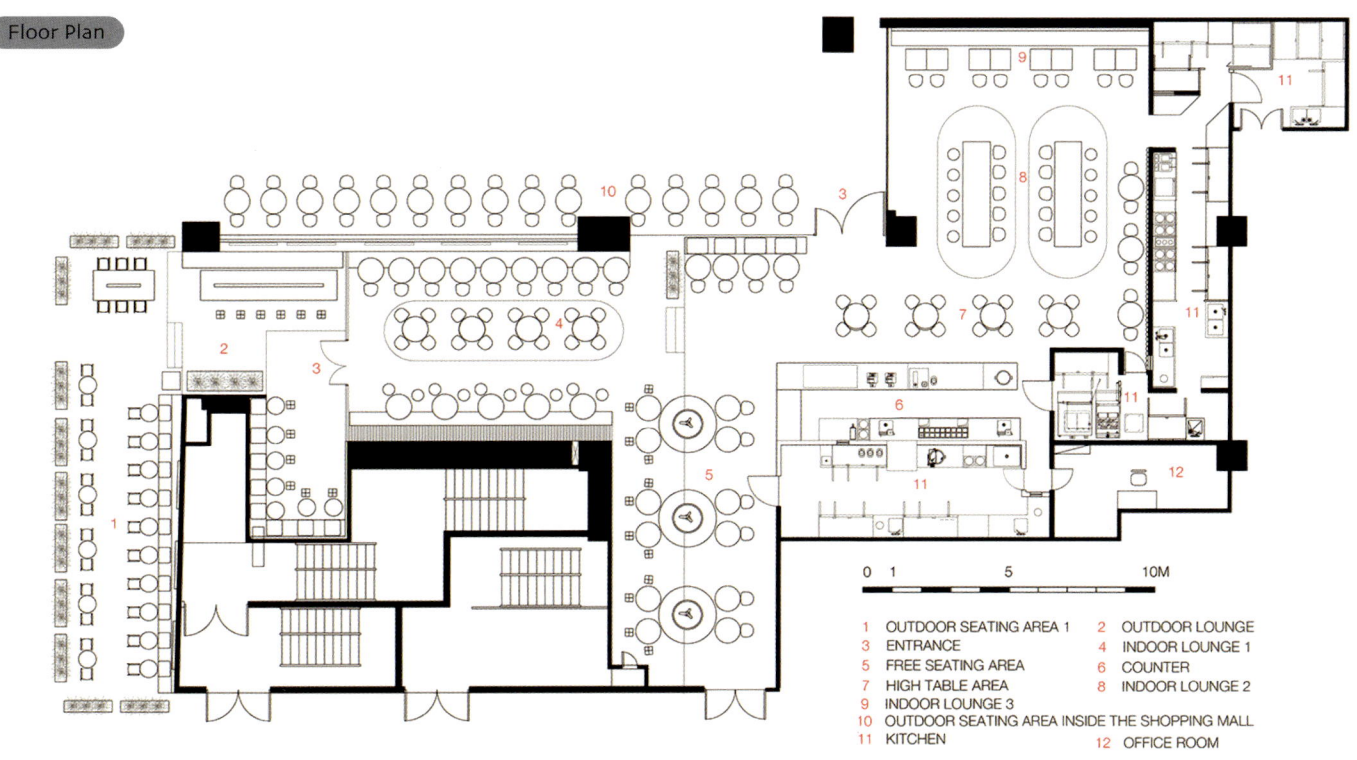

Floor Plan

0 1 5 10M
1 OUTDOOR SEATING AREA 1
2 OUTDOOR LOUNGE
3 ENTRANCE
4 INDOOR LOUNGE 1
5 FREE SEATING AREA
6 COUNTER
7 HIGH TABLE AREA
8 INDOOR LOUNGE 2
9 INDOOR LOUNGE 3
10 OUTDOOR SEATING AREA INSIDE THE SHOPPING MALL
11 KITCHEN
12 OFFICE ROOM

Large potted cacti deliver welcome pops of color and texture throughout the space, while geometric terrariums of succulents make for eye-catching tabletop details. Wall decorations – including circular mirrors, curved neon lights, and disc-shaped artworks – complement the arched lamella overhead, providing Gaga's digitally savvy clientele with an eminently shareable, social media-friendly backdrop. Aimed squarely at a young, trendy and sophisticated demographic of Chinese consumers, Gaga King Glory transports guests to a breezy and light oasis of calm. Conceived as an idyllic beach – just a short distance from Shenzhen's industrious port – this newest venue maintains the brand's reputation for stylish design, exceptional coffee, and above all, as a place to inspire.

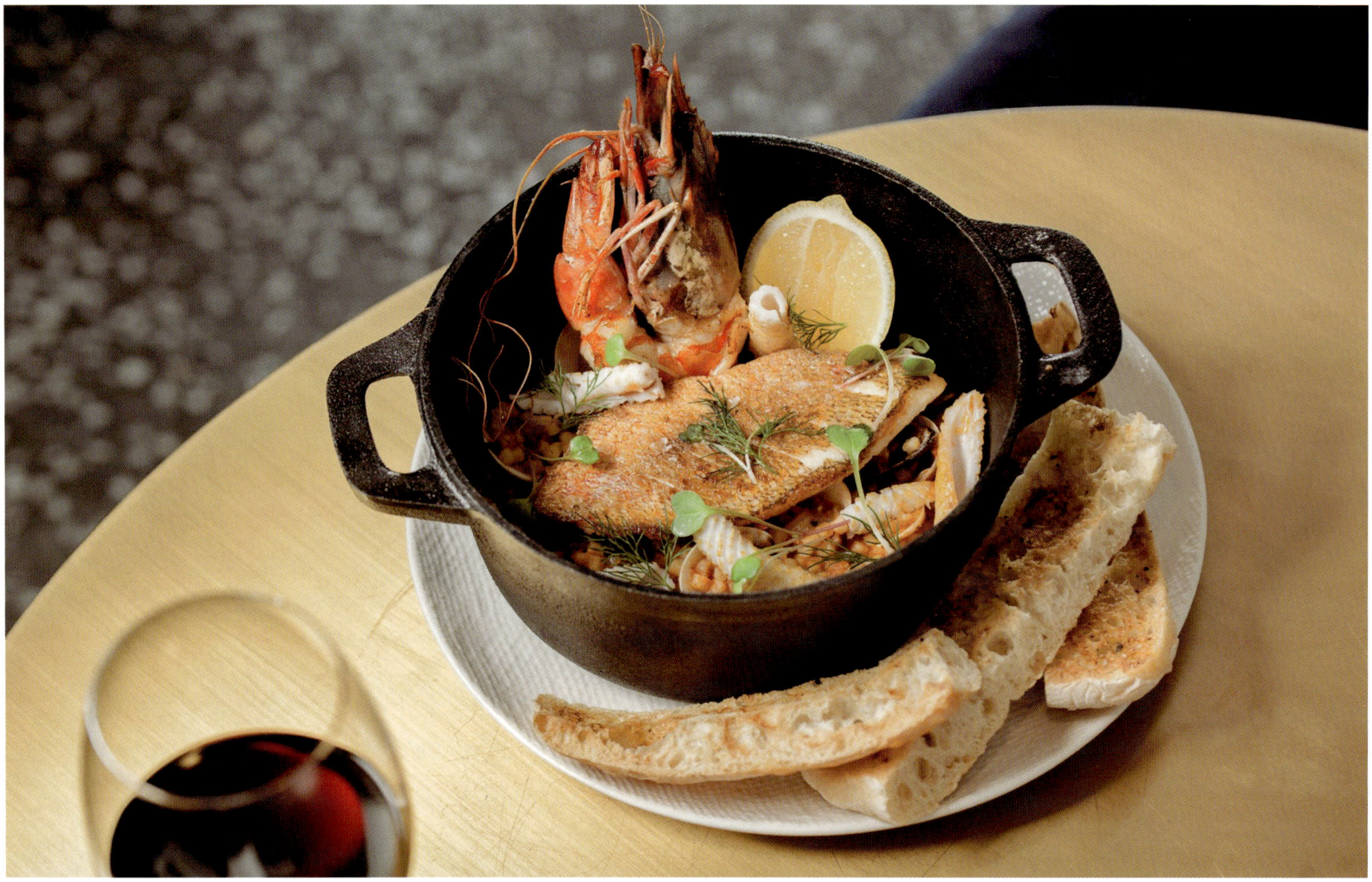

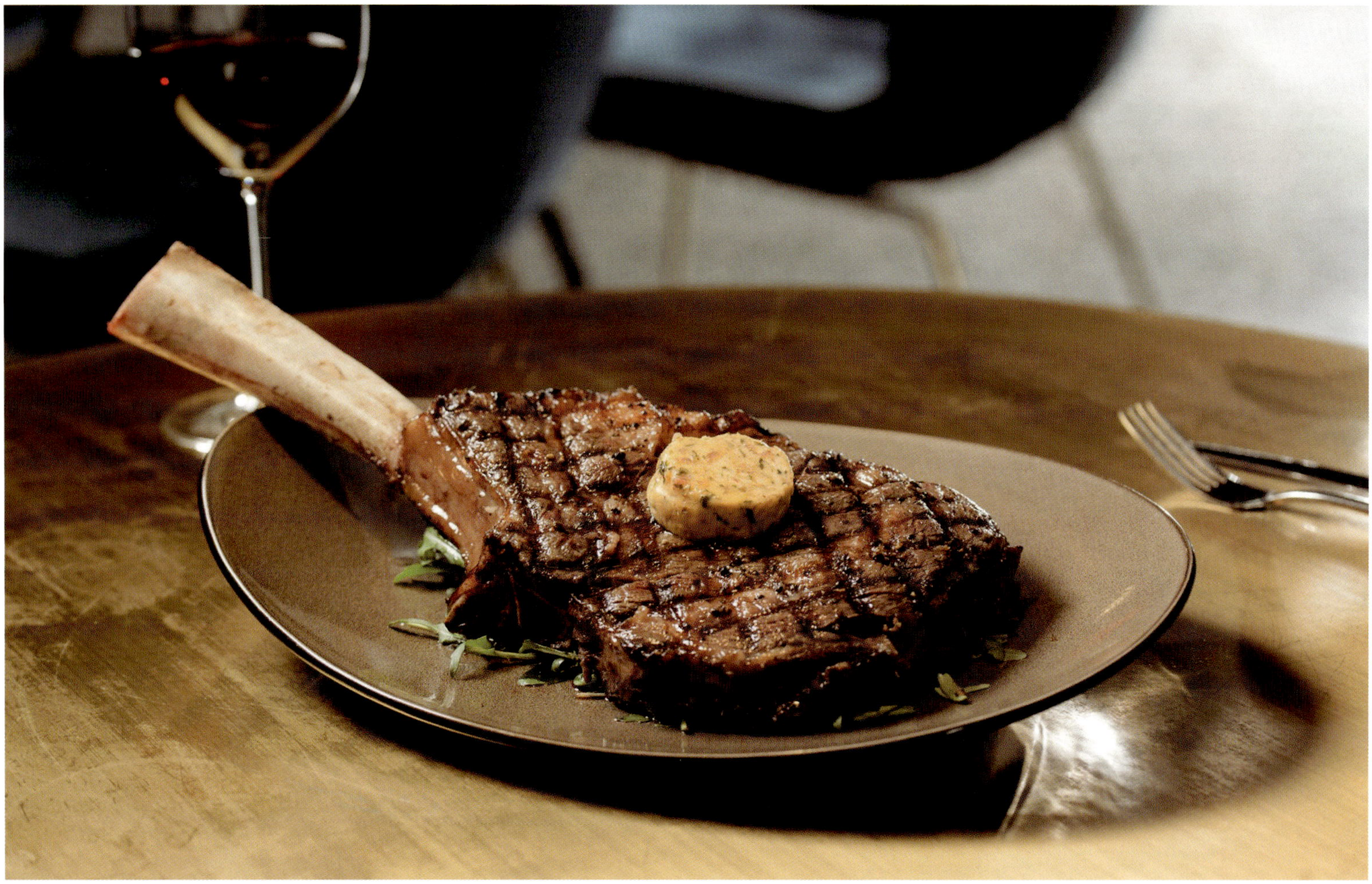

Emem Design?

EMEM DESIGN is an interior design studio based in Ho Chi Minh City, which specializes in residential, apartment, retail, hospitality, and commercial practices. We provide an all-in-one interior design solution that starts from concept development to the final construction.

Design: Emem Design   **Homepage:** www.behance.net/ememdesignvn   **Location:** Ho Chi Minh City, Vietnam
Area: 700 sq.m   **Photographer:** lilo studio

Vòm Coffee is the latest in a series of coffee store projects completed by Emem Design in 2020. Inspired by the warm neutral palette of the Mediterranean landscape, Emem keeps both paint and furnishings minimal while warming up space with wood and natural materials, like trees, stone and terrazzo with geometric patterns. It's really a play of inside and outside living, bathed in warmth.

Dom Coffee is a dating place, a working corner, a place to gather friends after stressful working and studying hours. The Dome space with Mediterranean style and refreshing drinks is like an oasis in the heart of the crowded city where you choose to stop to cool off, to find relaxation before continuing your journey. Going to many places, I know where I like, When you meet many people, you will never know who you love...Let's visit the dome space and enjoy a little familiar and pleasant atmosphere.

"The stream of people rushes through" in front of the porch of the Dome. The weather after the rain leaves a very pleasant new day, friends, let's go on a date to see the streets of Saigon and enjoy a cool drink from Vom.

Vom Coffee District 10 is the second house of Khom Coffee, which has just been launched with a new name "Voc" and has a lot of virtual living corners.
Vom Coffee in District 10 has a Mediterranean style with white tones combined with wood, bamboo and rattan creating a very familiar feeling. With a 4-storey design, the space here is quite large, designed in a modern style, with many views to check-in. My favorite is the terrace and I'm sure this will also be the point to attract people here because the stairs to take pictures with the sky are so beautiful.

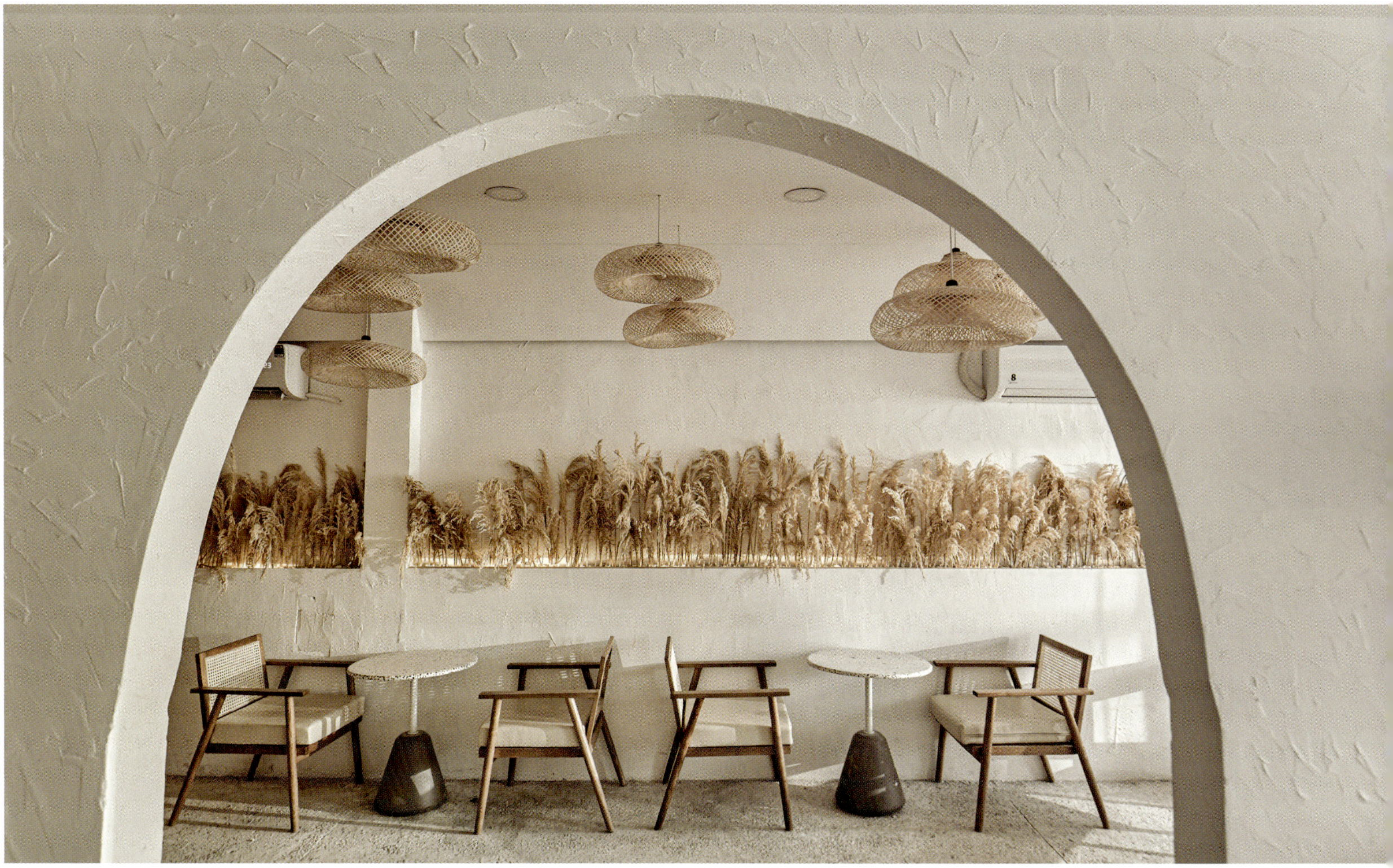

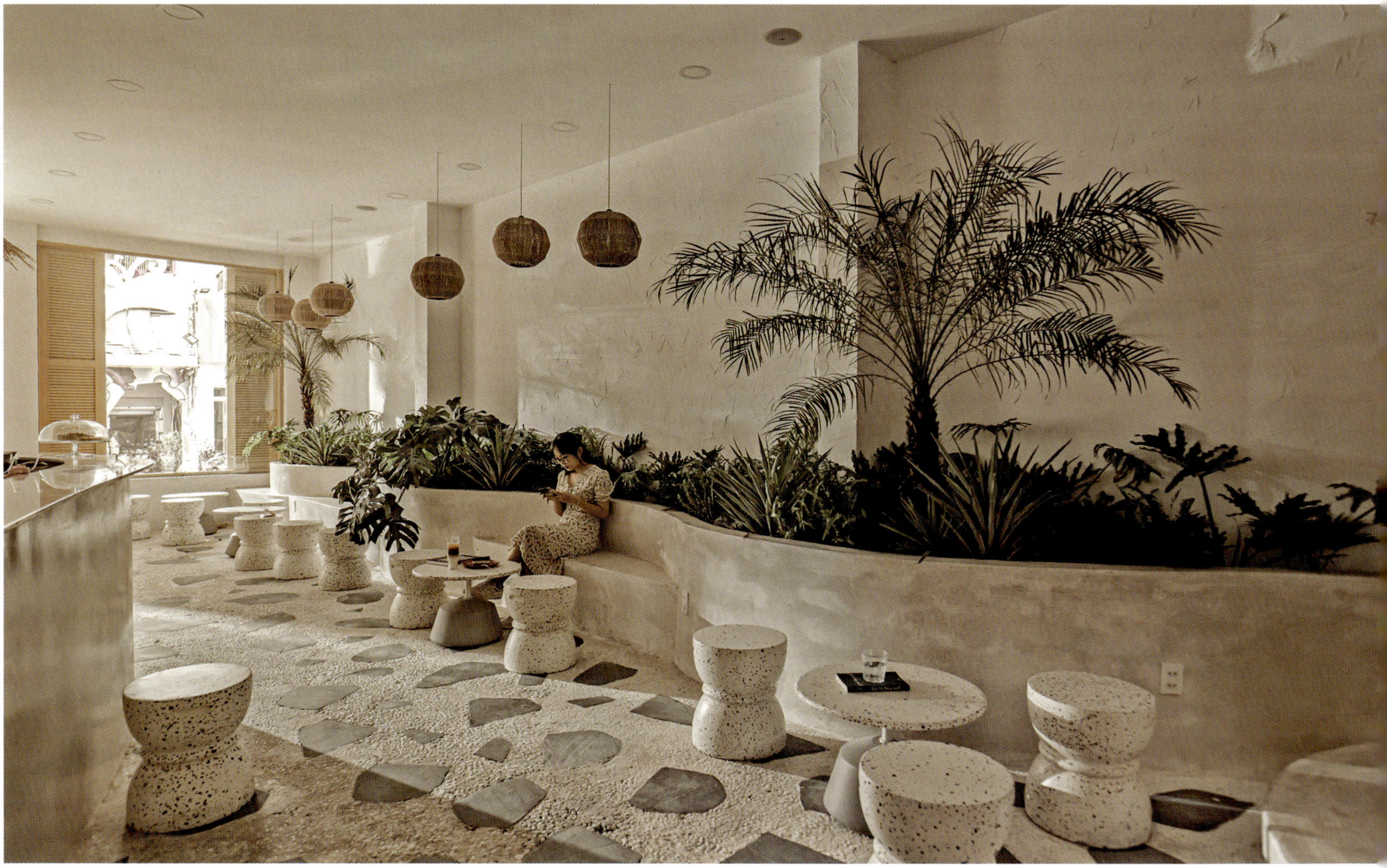

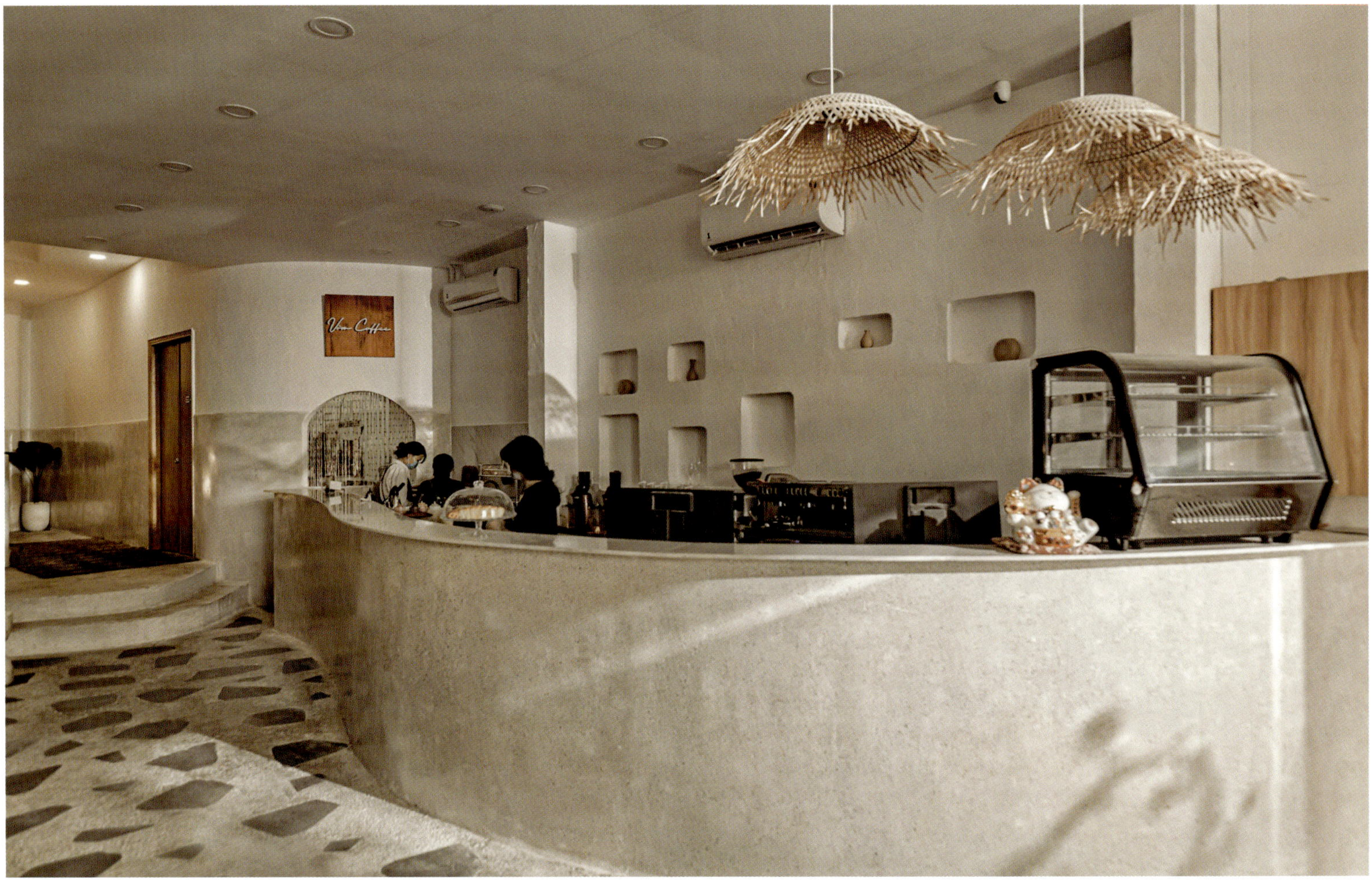

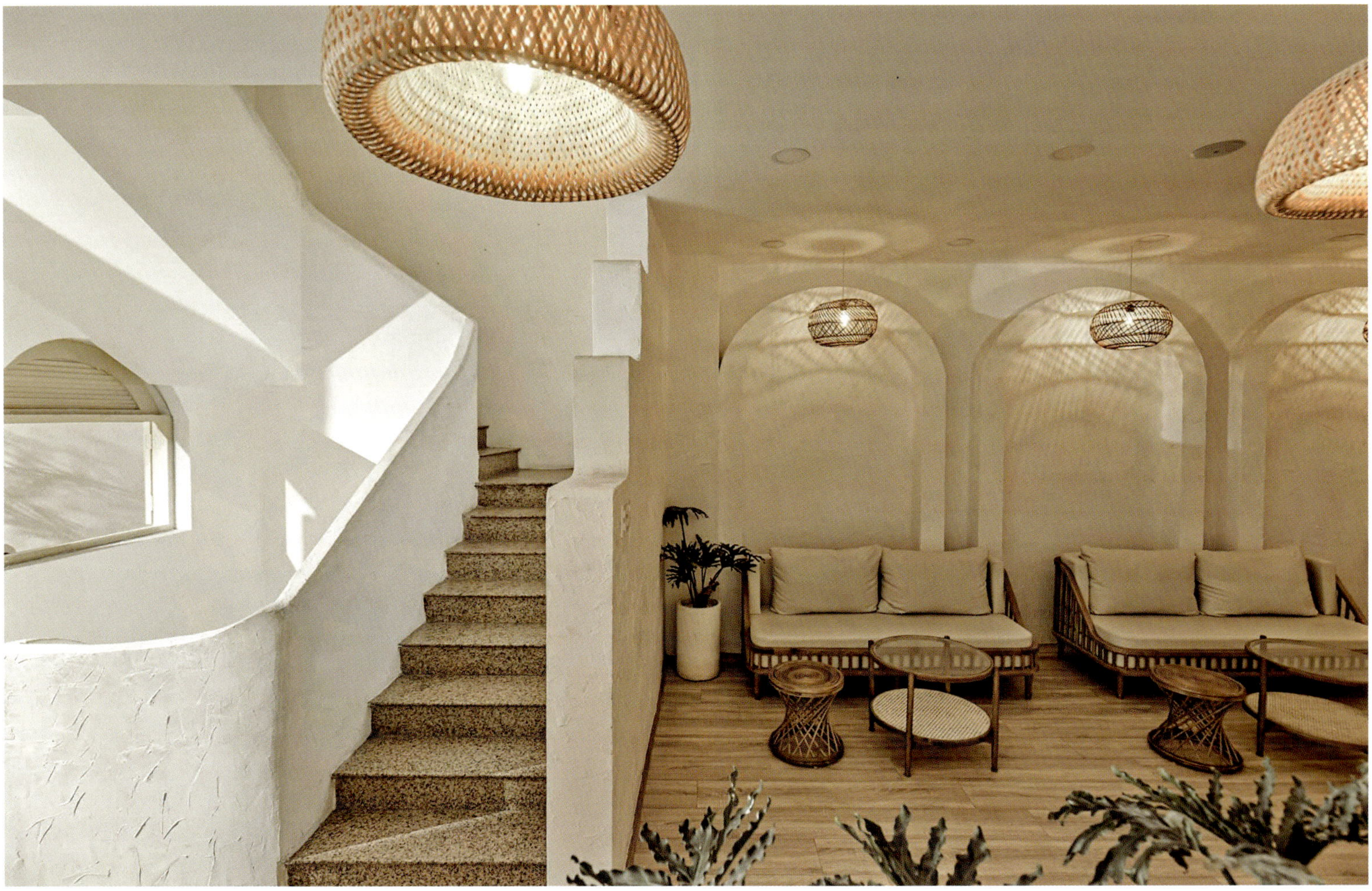

203

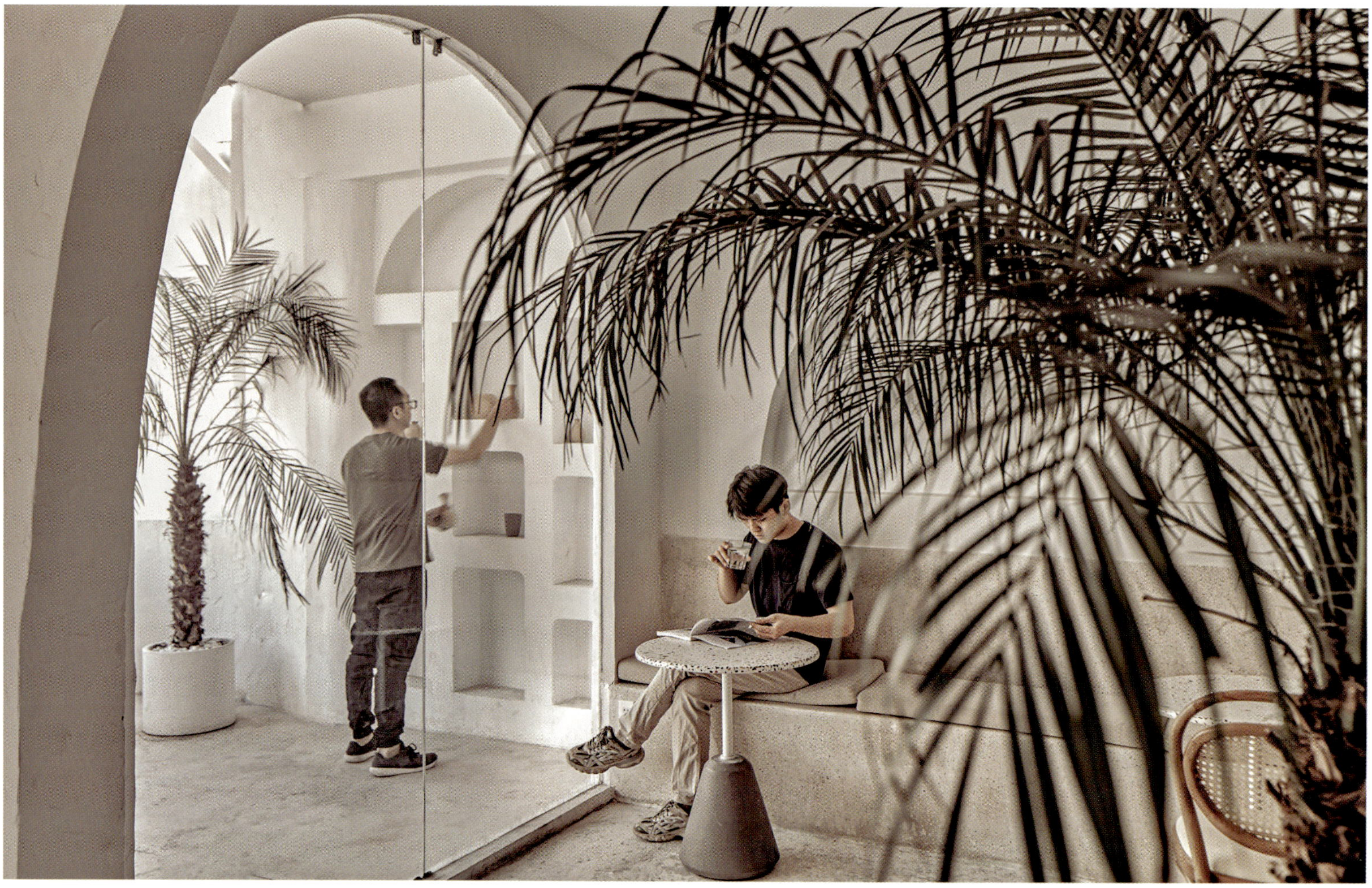

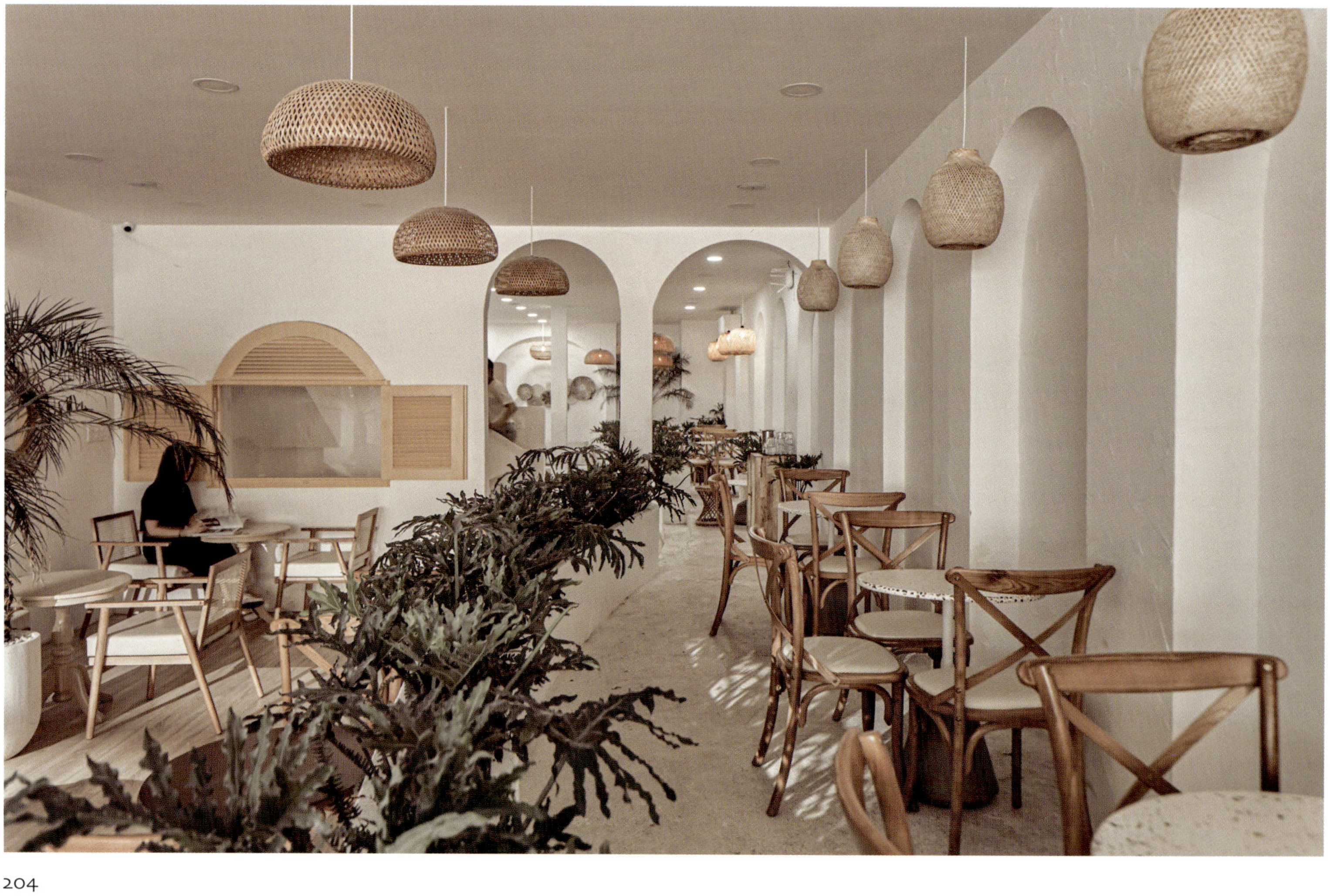

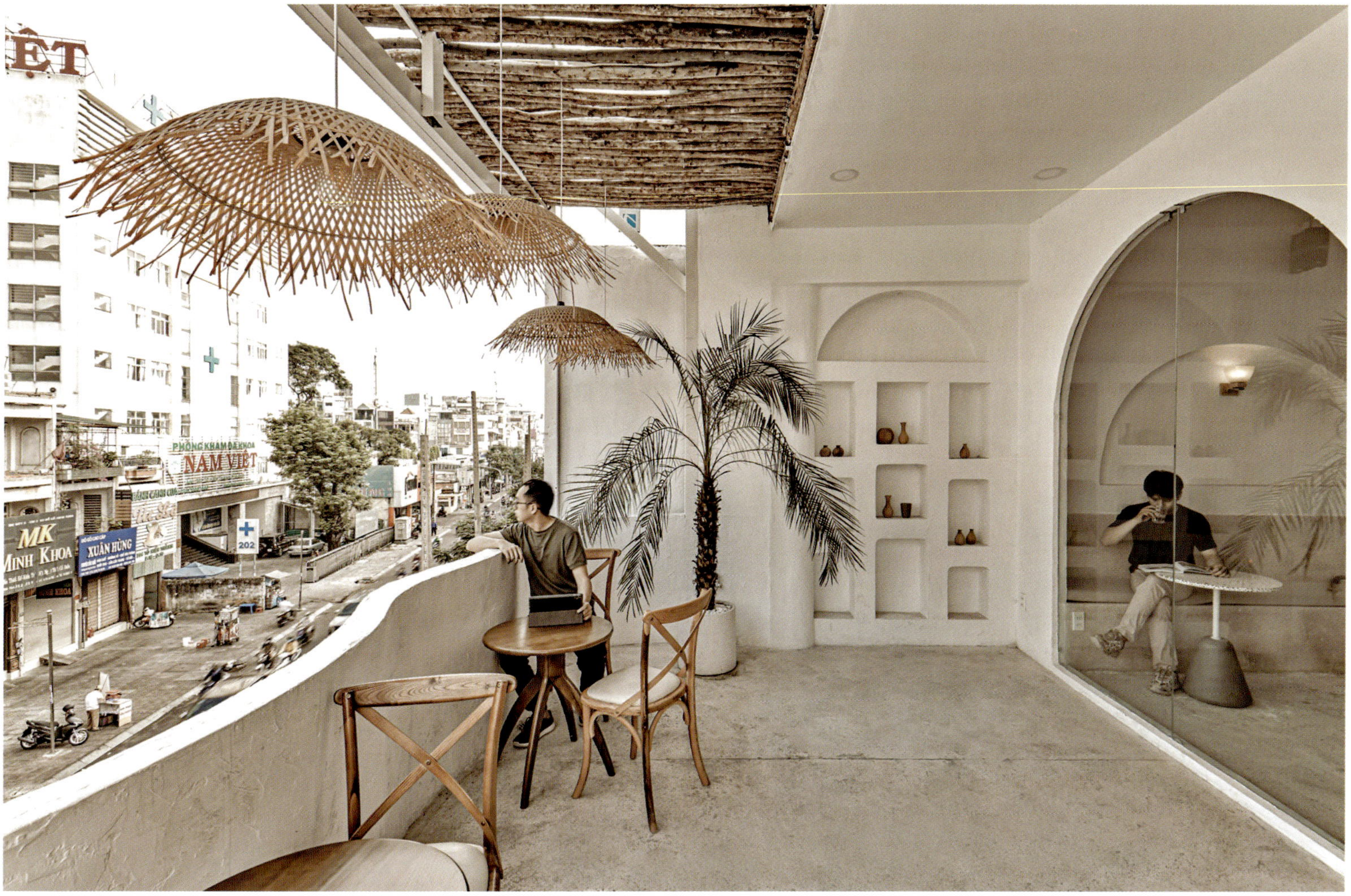
MK
MINH KHOA
XUÂN HÙNG
NAM VIỆT

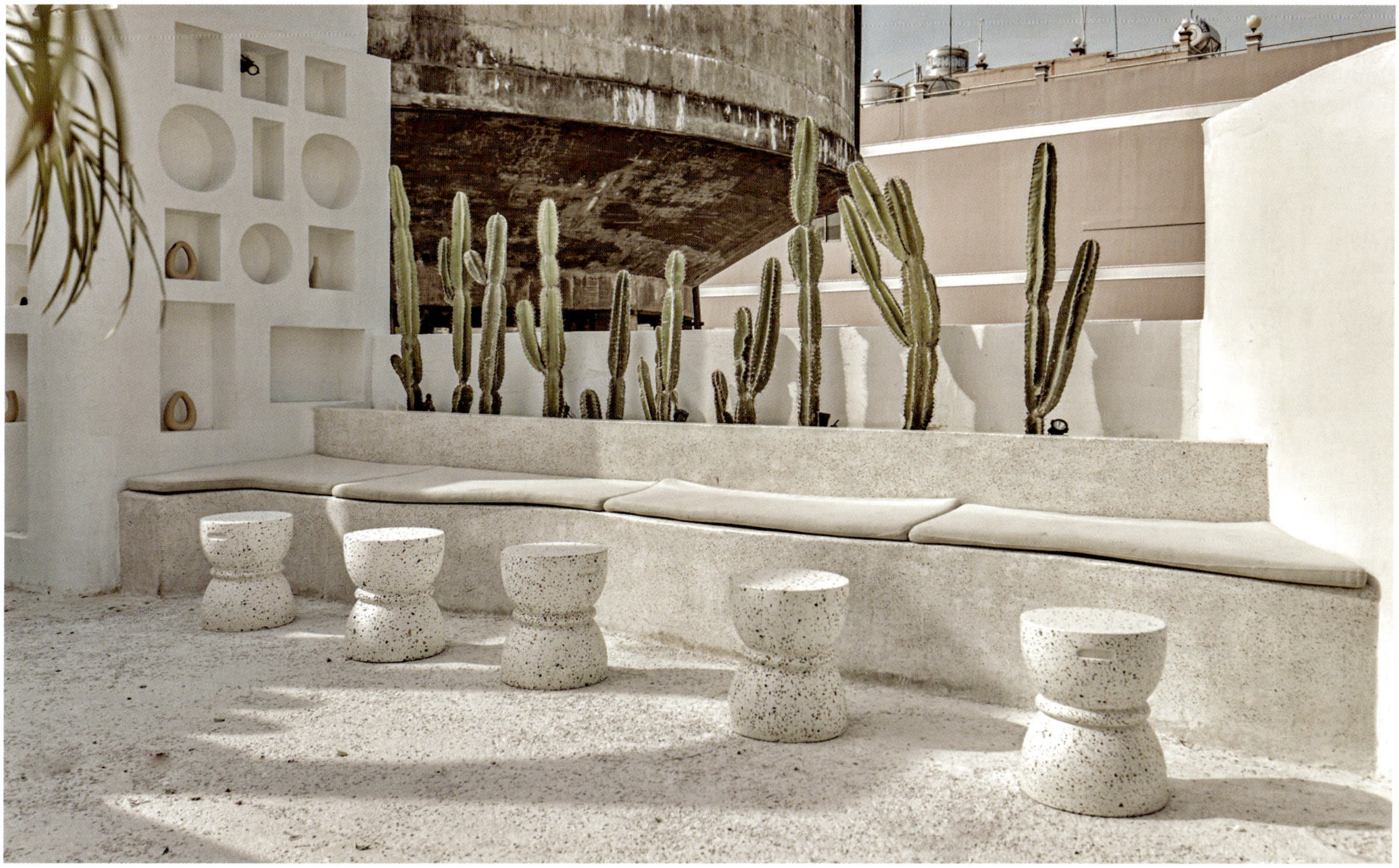

Pano Brot & Kaffee
Stuttgart, Germany

PANO®
BROT & KAFFEE

DIA-Dittel Architekten?

In our architectural office, architects, interior architects and communication designers work together to make unique creations a reality. For every project, we go through an exciting creation process that is subject to our quality standards and that, in symbiosis with inspiration, well-grounded knowledge and teamwork based on trust, results in successful project implementation.

Design: *DIA-Dittel Architekten* **Homepage:** *https://di-a.de* **Location:** *Stuttgart, Germany* **Area:** *225 m²* **Photographs:** *Dittel Architekten GMBH*

Unwind. Feel comfortable. Have a snack. In the heart of Stuttgart, DIA has created the town's first PANO branch at the new Gerber shopping centre. The café invites shoppers to enjoy home-made products from organic and local farms in a cosy setting. Visitors experience a warm, relaxed atmosphere and are invited into the living room like guests calling on good friends.

An extraordinary ceiling element, which unfolds from the entrance through to the rear section, leads into the 225m² room. The use of high-quality materials, such as solid oak, hand-stitched real leather and Italian clay flagstones, represents corporate values such as tradition, honesty and quality. In this way, the PANO brand can be comprehended in a holistic way.

The solid oak table, which is over 5m long and 1.15m wide, is the distinctive feature of the room, complemented by an impressive bookshelf over 5m high. The latter sets the tone for the dining area with approx. 92 seats in its characteristic openness, creating an exclusive area in which to present the diverse assortment of products.

A wide variety of seating options offers the very varied target group the right seat for individual requirements. Whether you are after a quick snack 'à deux' with a view across the Gerber quarter or a good old chat with friends over coffee in front of a cosy fire, the interior decoration offers the ideal opportunity for enjoying a break from your daily routine to the fullest.

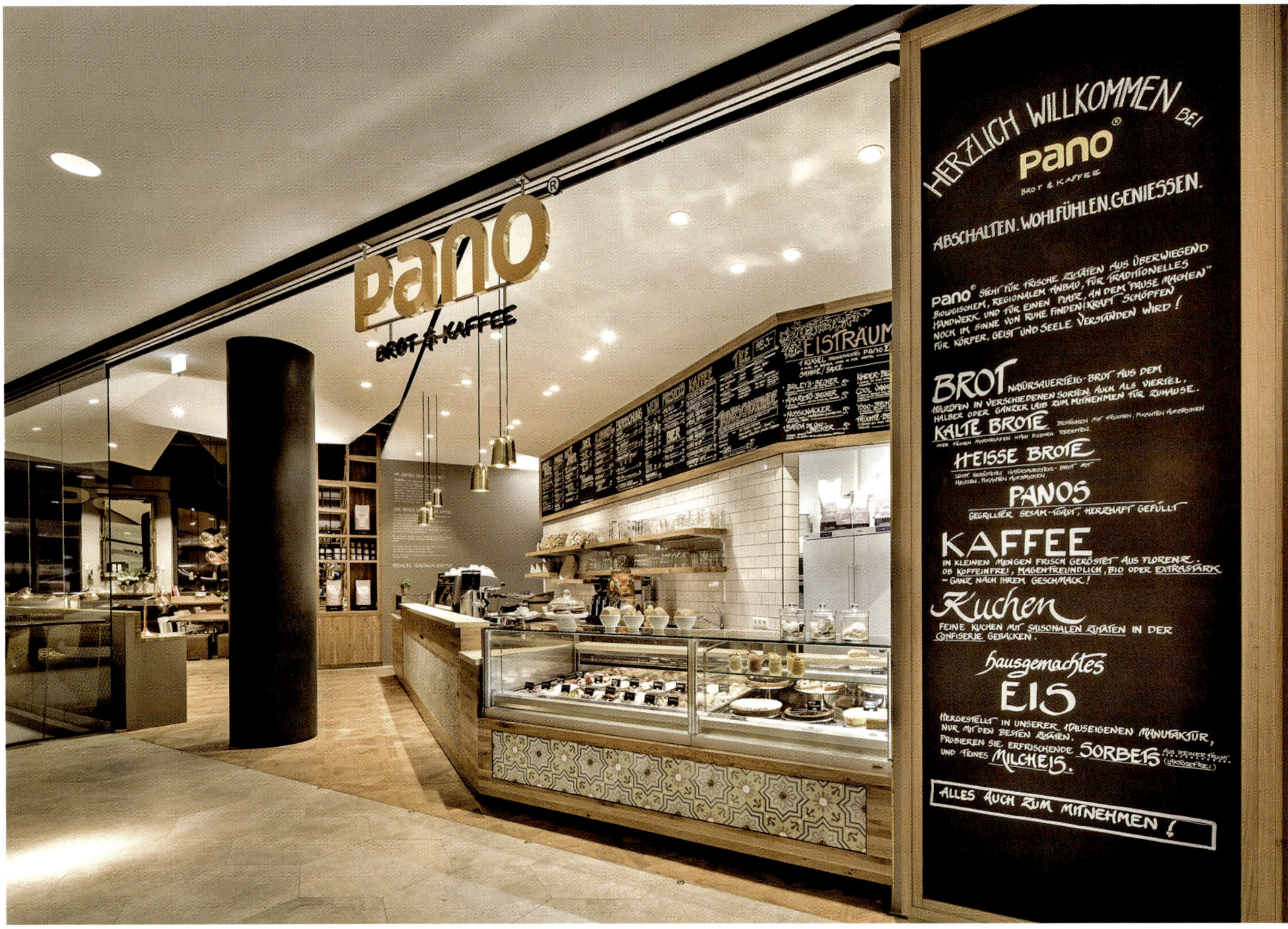
pano
BROT & KAFFEE
EISTRAUM
HERZLICH WILLKOMMEN BEI
pano
BROT & KAFFEE
ABSCHALTEN. WOHLFÜHLEN. GENIESSEN.
pano STEHT FÜR FRISCHE ZUTATEN AUS ÜBERWIEGEND BIOLOGISCHEM, REGIONALEM ANBAU, FÜR TRADITIONELLES HANDWERK UND FÜR EINEN PLATZ, AN DEM "PAUSE MACHEN" NOCH IM SINNE VON RUHE FINDEN! KRAFT SCHÖPFEN FÜR KÖRPER, GEIST UND SEELE VERSTANDEN WIRD!
BROT
NATURSAUERTEIG-BROT AUS DEM HOLZOFEN IN VERSCHIEDENEN SORTEN. AUCH ALS VIERTEL, HALBER ODER GANZER LAIB ZUM MITNEHMEN FÜR ZUHAUSE.
KALTE BROTE
HEISSE BROTE
PANOS
GEGRILLTER SESAM-TOAST, HERZHAFT GEFÜLLT
KAFFEE
IN KLEINEN MENGEN FRISCH GERÖSTET AUS FLORENZ. OB KOFFEINFREI, MAGENFREUNDLICH, BIO ODER EXTRASTARK - GANZ NACH IHREM GESCHMACK!
Kuchen
FEINE KUCHEN MIT SAISONALEN ZUTATEN IN DER CONFISERIE GEBACKEN.
hausgemachtes
EIS
HERGESTELLT IN UNSERER HAUSEIGENEN MANUFAKTUR, NUR MIT DEN BESTEN ZUTATEN. PROBIEREN SIE ERFRISCHENDE SORBETS UND FEINES MILCHEIS.
ALLES AUCH ZUM MITNEHMEN!

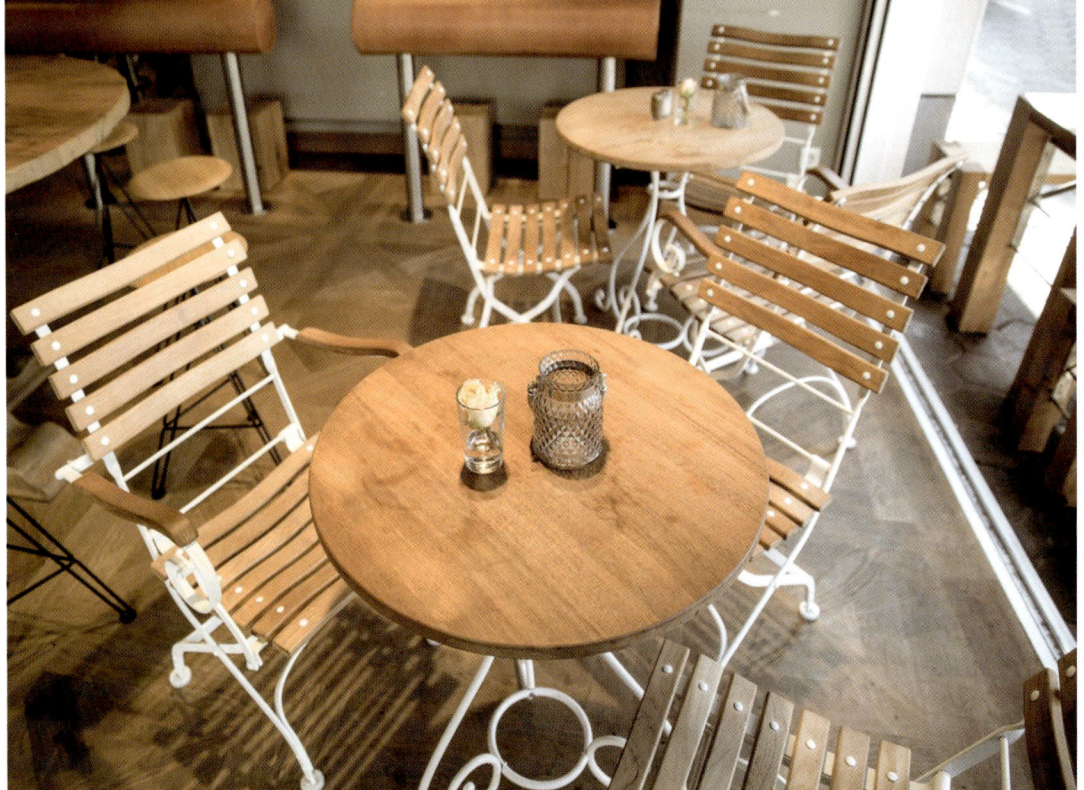

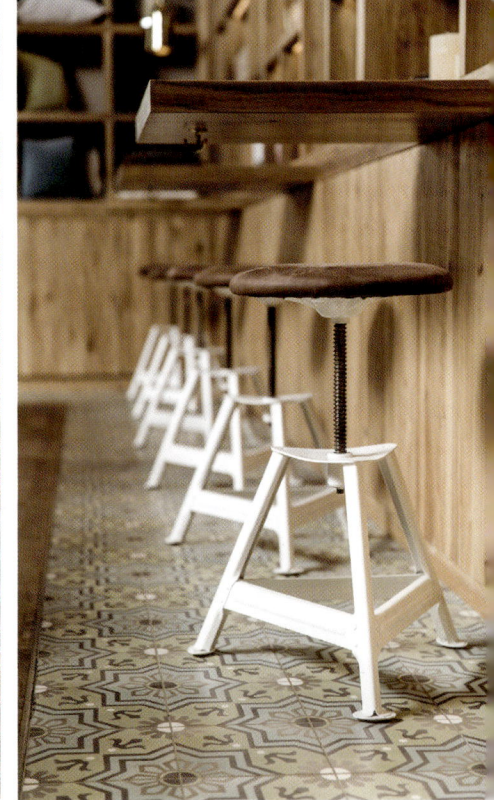

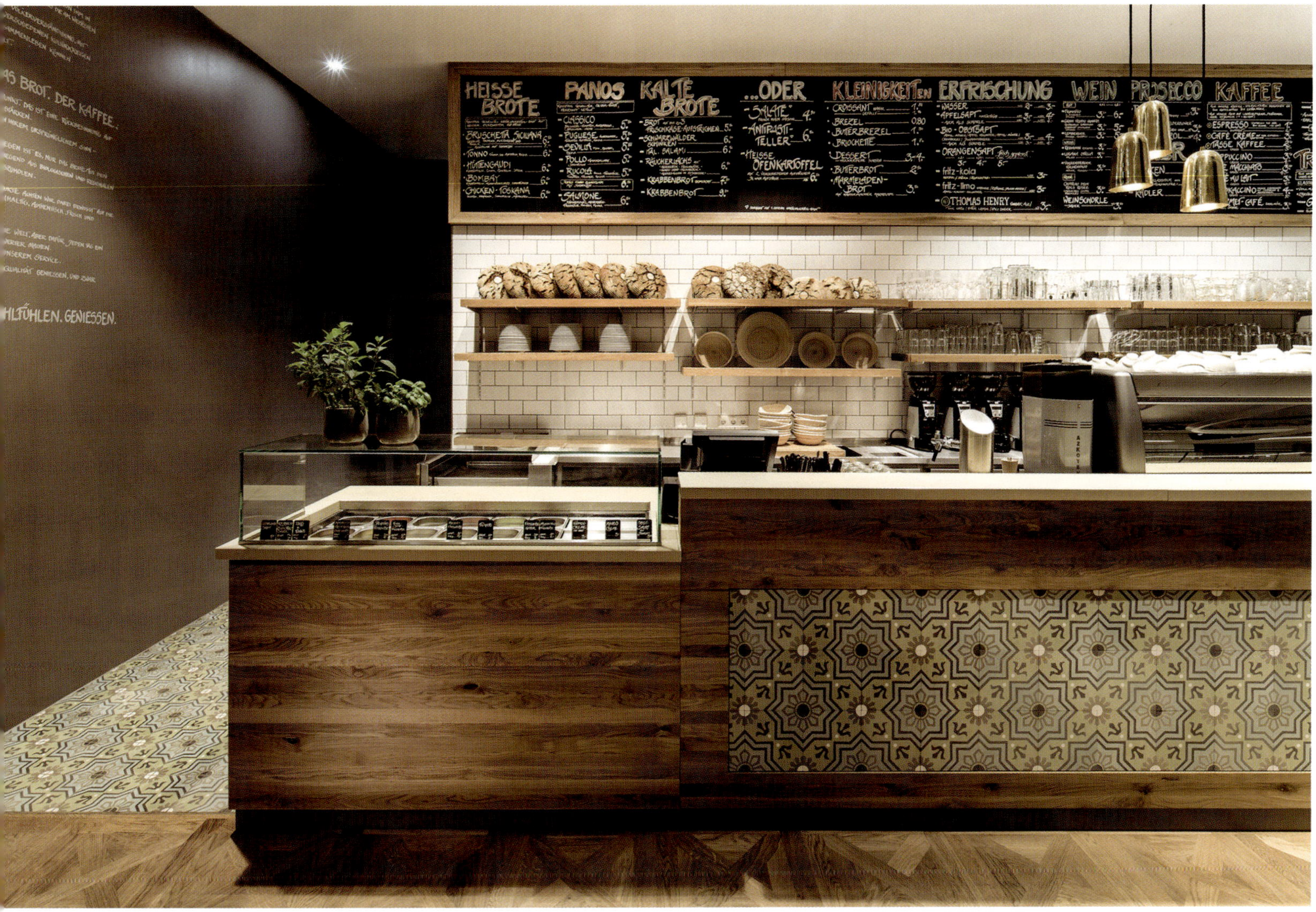
HEISSE BROTE
PANOS
KALTE BROTE
...ODER
KLEINIGKEITEN
ERFRISCHUNG
WEIN
PROSECCO
KAFFEE

PANOS
CLASSICO
PUGLIESE
SEVILLA
POLLO
RUCOLA
SALMONE

pano
BROT & KAFFEE

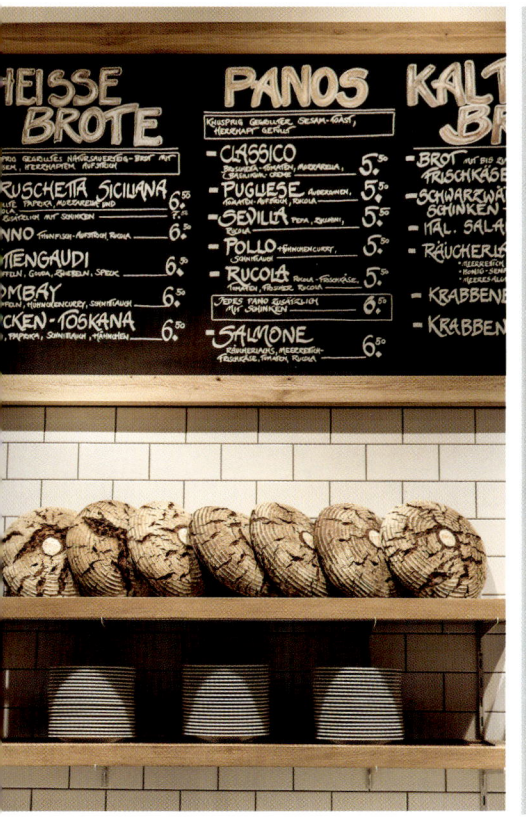

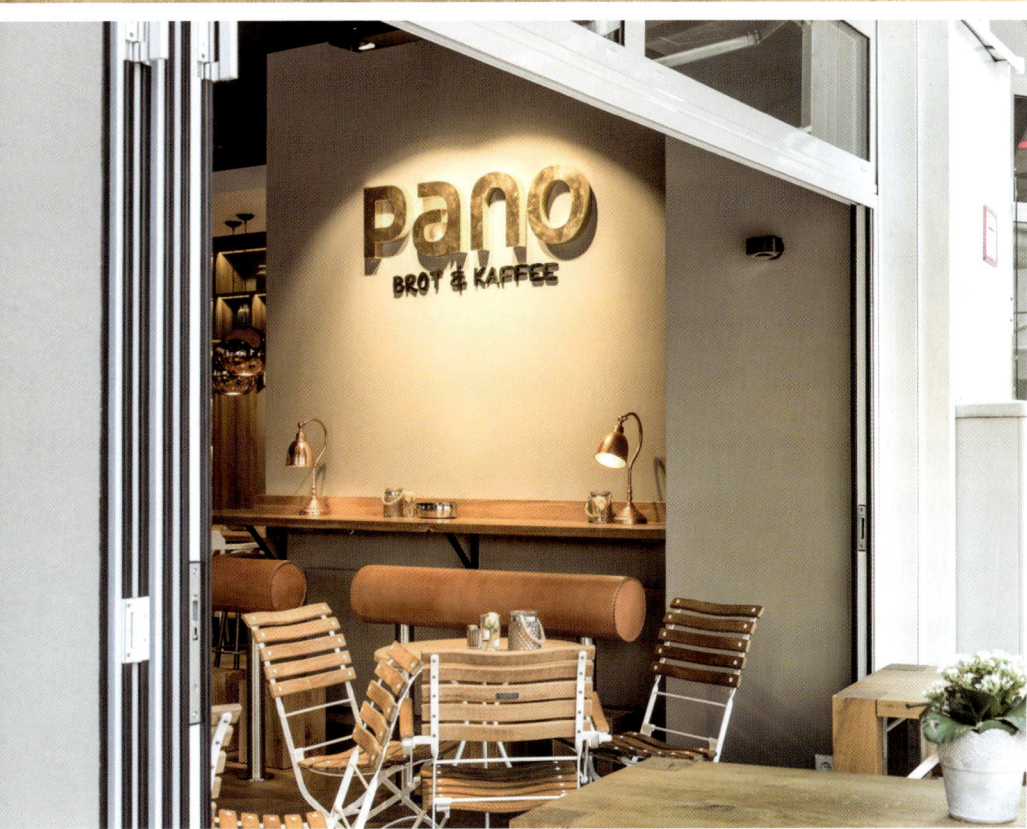

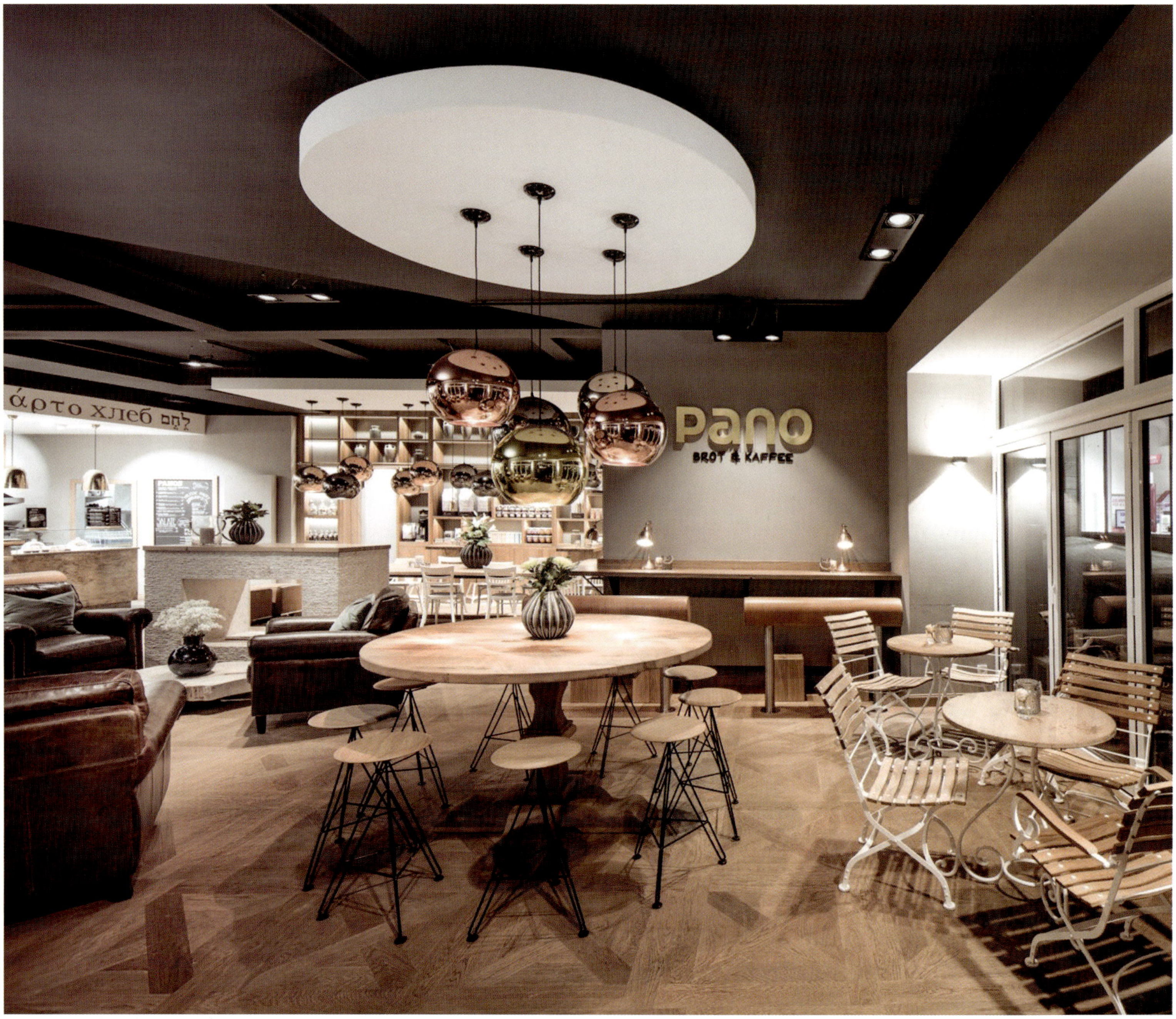
άρτο хлеб לחם
Pano
BROT & KAFFEE

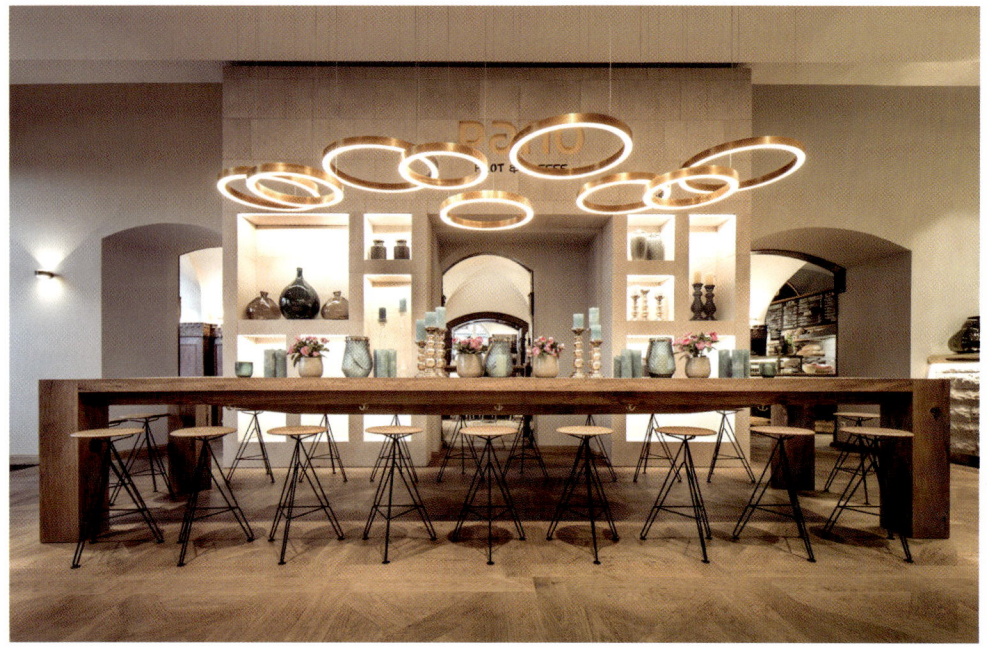

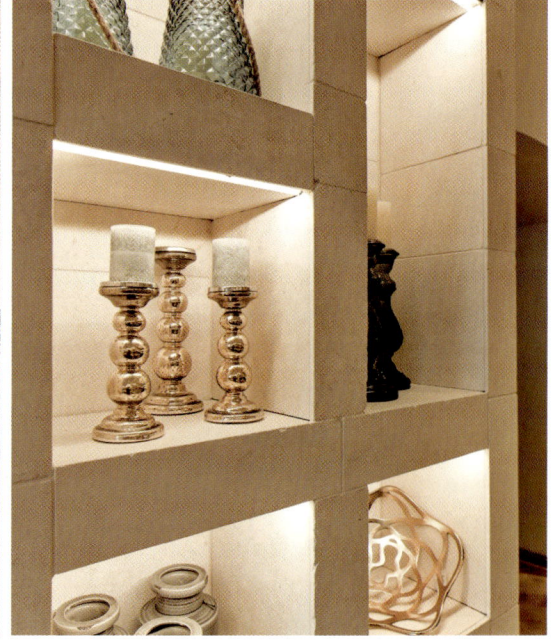

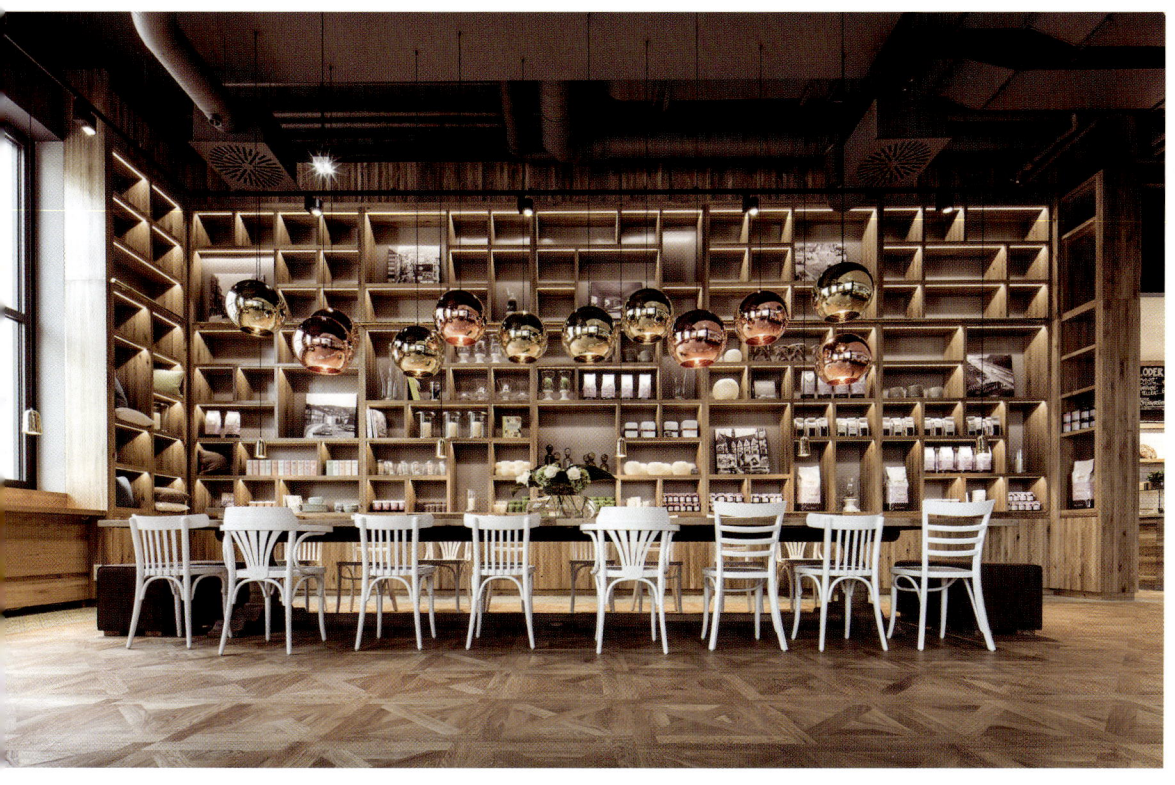

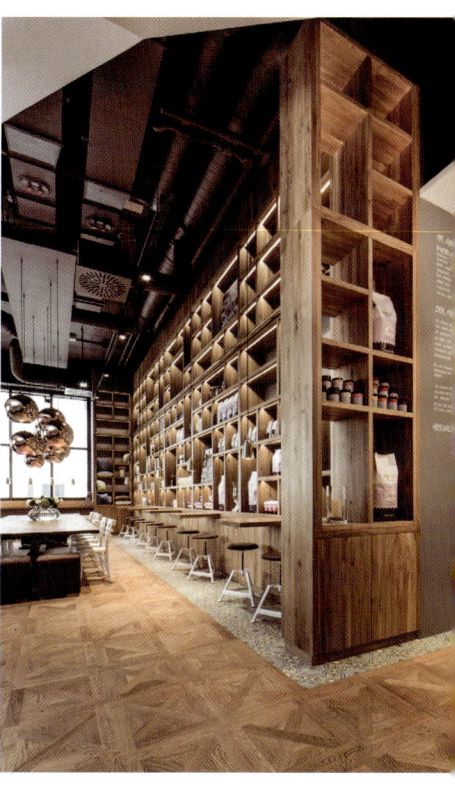

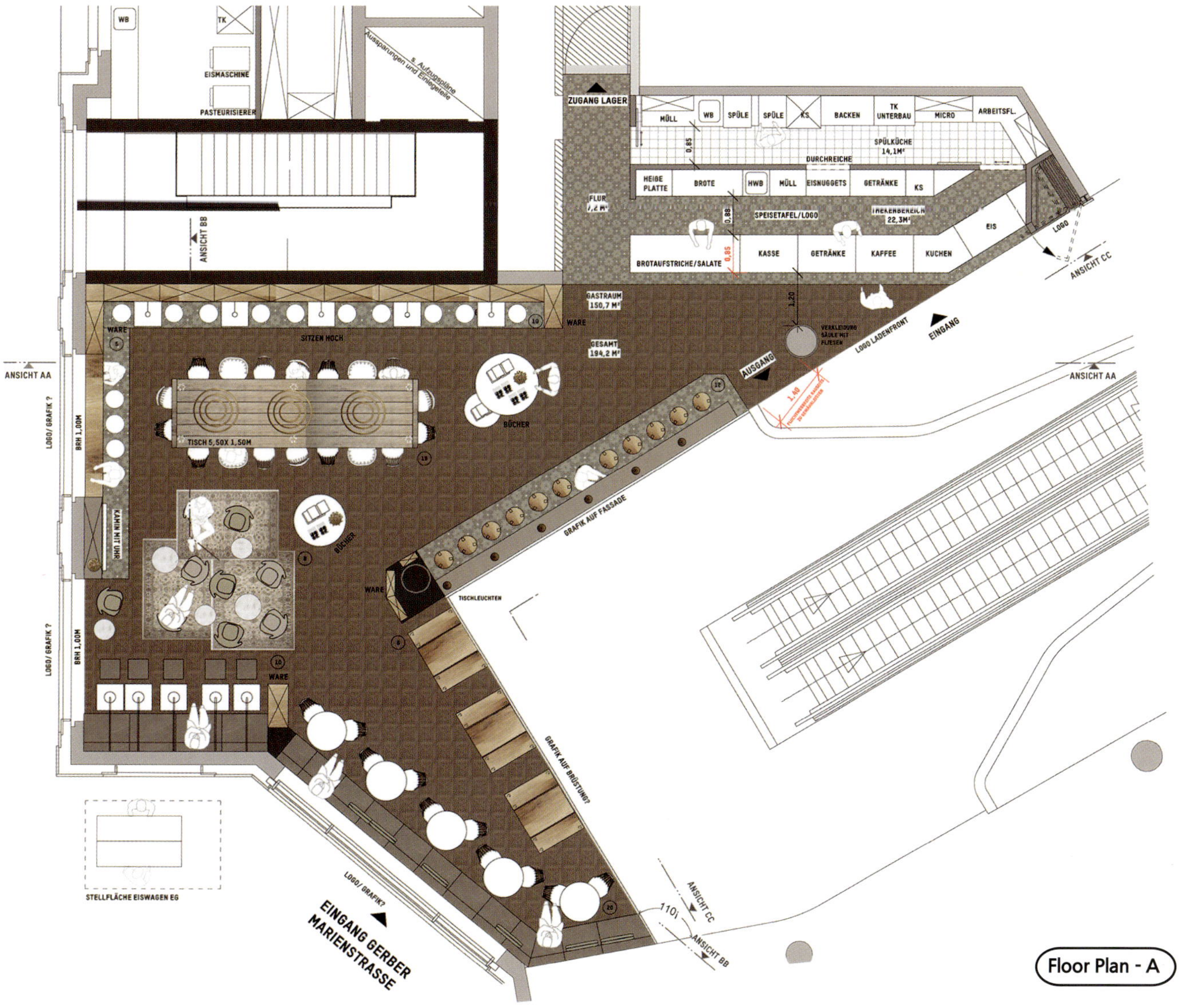

WB
TK
EISMASCHINE
PASTEURISIERER
ZUGANG LAGER
MÜLL
WB
SPÜLE
SPÜLE
KS
BACKEN
TK UNTERBAU
MICRO
ARBEITSFL.
SPÜLKÜCHE 14,1M²
DURCHREICHE
HEIßE PLATTE
BROTE
HWB
MÜLL
EISNUGGETS
GETRÄNKE
KS
SPEISETAFEL/LOGO
EIS
BROTAUFSTRICHE/SALATE
KASSE
GETRÄNKE
KAFFEE
KUCHEN
LOGO
ANSICHT CC
ANSICHT BB
GASTRAUM 150,7 M²
GESAMT 194,2 M²
WARE
SITZEN HOCH
EINGANG
AUSGANG
LOGO LADENFRONT
ANSICHT AA
ANSICHT AA
LOGO/ GRAFIK ?
BRH 1,00M
TISCH 5,50X 1,50M
BÜCHER
GRAFIK AUF FASSADE
TISCHLEUCHTEN
STELLFLÄCHE EISWAGEN EG
LOGO/ GRAFIK?
EINGANG GERBER MARIENSTRASSE
GRAFIK AUF BRÜSTUNG?
ANSICHT CC
ANSICHT BB

Floor Plan - A

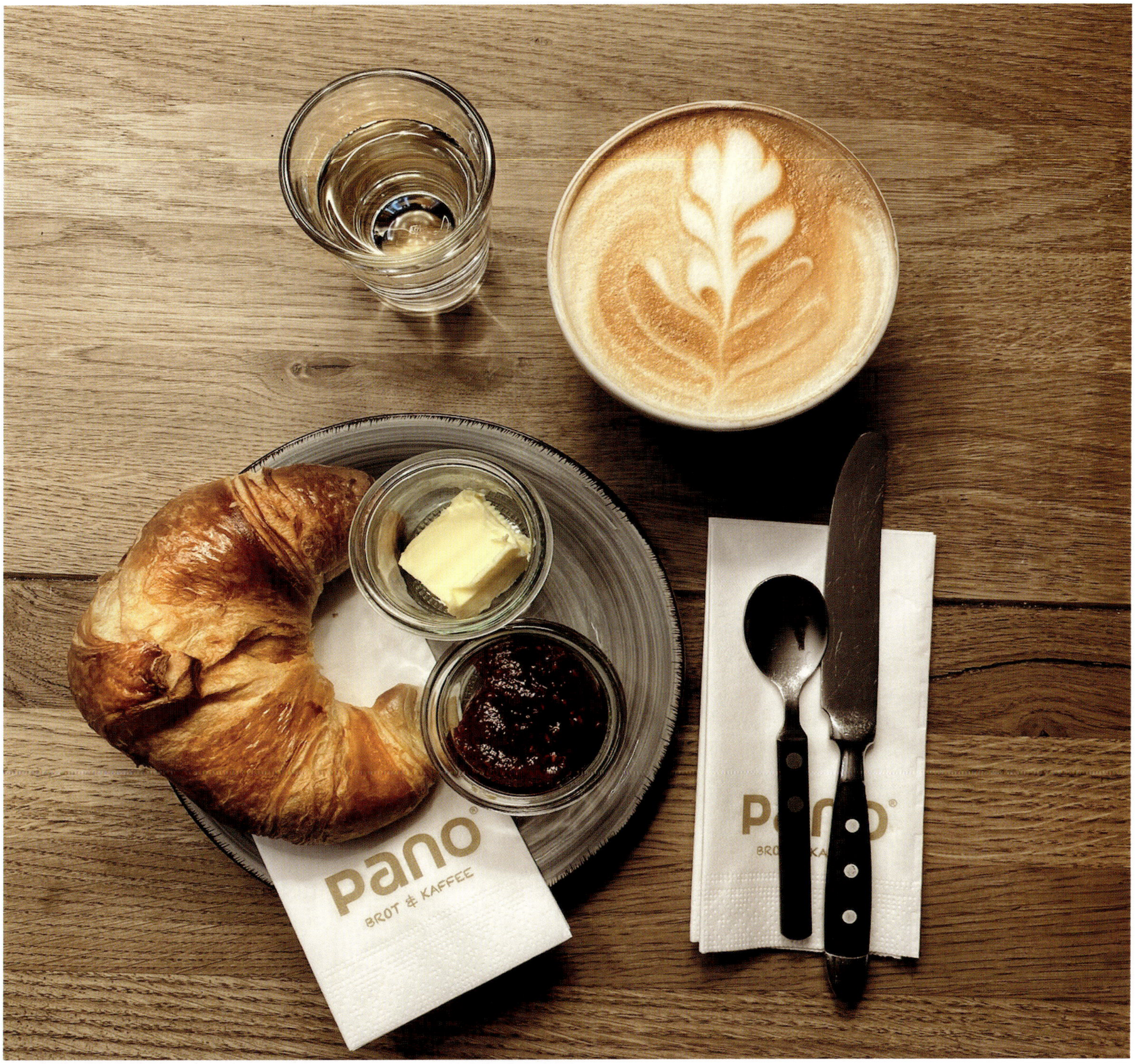
pano
BROT & KAFFEE

pano

pano
KAFFEE
Pura
pano
KAFFEE
Pura

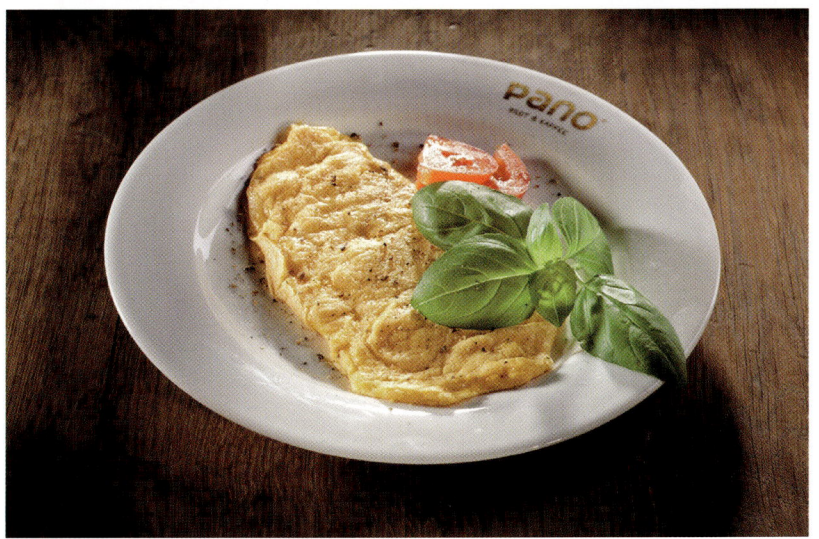

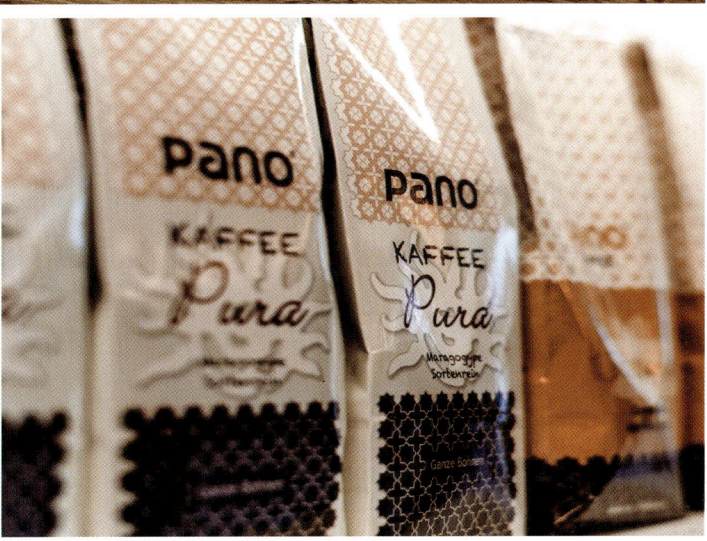

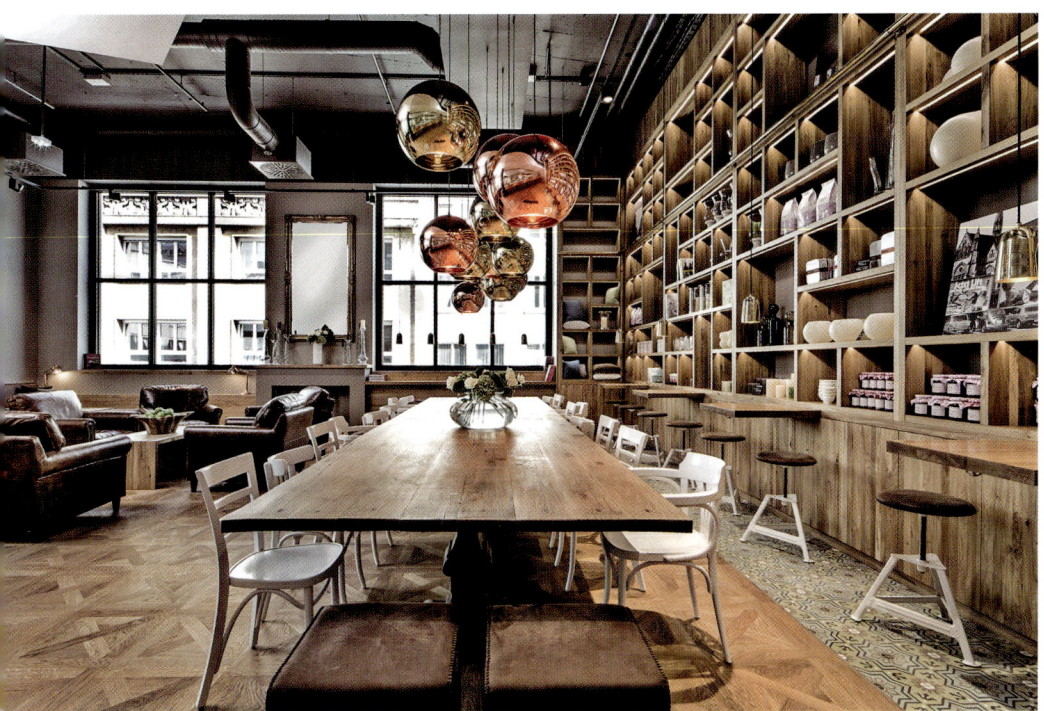

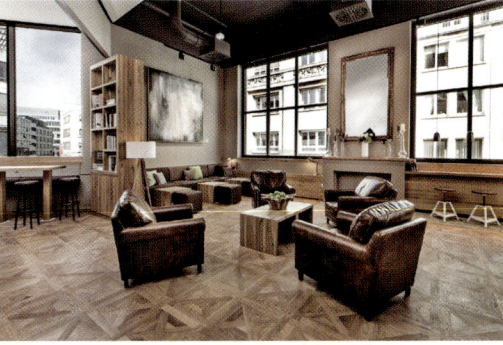

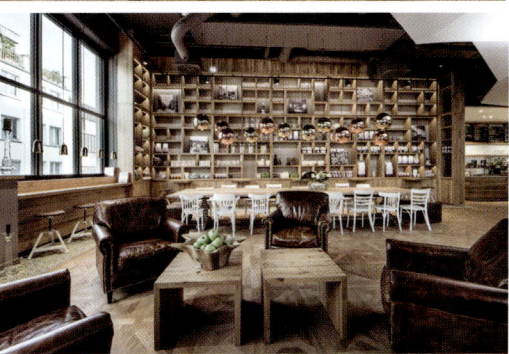

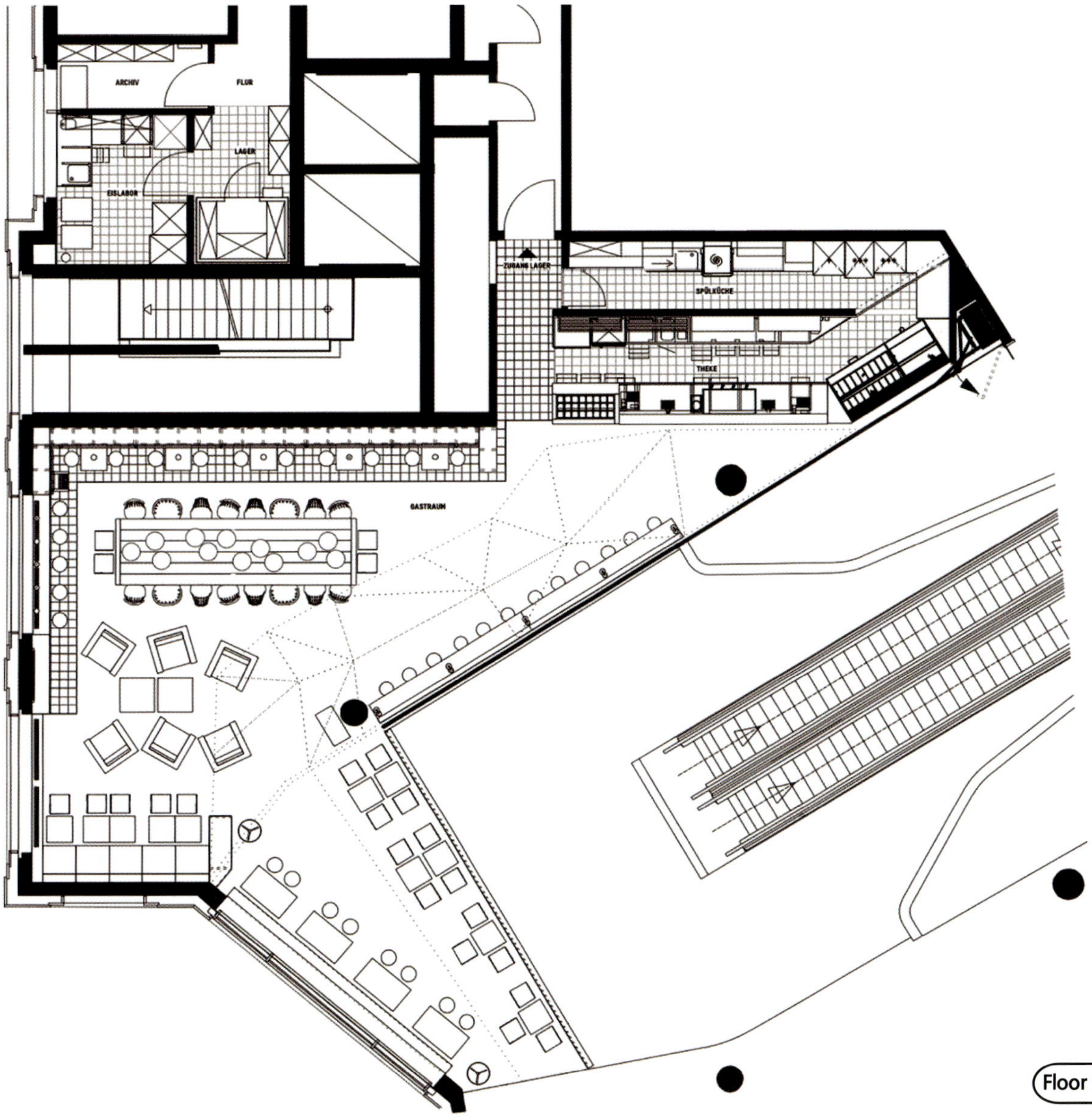
ARCHIV
FLUR
LAGER
EISLABOR
ZUGANG LAGER
SPÜLKÜCHE
THEKE
GASTRAUM

Floor Plan - A'

NANA Coffee Roasters
Bangna, Thailand

IDIN Architects?

An acronym for Integrating Design Into Nature, IDIN Architects was founded in 2004. We perceive 'nature' in two ways. Firstly, nature can be defined as the ecology around us. Secondly, it can also refer to different mannerisms and personalities. The design philosophy and attention of IDIN are to merge this sense of surroundings, the 'natures', to the architectural aesthetic. This merge is done through a process of analyzing and prioritizing the different needs and requirements of each project.

In addition to being an acronym, the Thai word "idin" is used to describe the natural phenomenon when soil releases a beautiful scent after rainfall. This symbolizes Thailand's tropical climate, something all IDIN designs aim to respond to. Our emphasis is therefore placed not only on aesthetics, but also on being practical, in order to suit this temperate environment.

Design: IDIN Architects   **Homepage:** http://www.idin-architects.com   **Landscape design:** Terrains + Open Space   **Location:** Bangna, Thailand   **Area:** 250 m²   **Photographs:** W-Workspace

---

The concept of NANA Coffee Roasters, Bangna Branch is to create spaces that enhance the coffee-drinking experience where the coffee becomes the main center of attention. The architectural expression of the project was simplified, while the design is shifted to combine architectural spaces harmoniously with the landscape to create a lush atmosphere that draws the visitors away from the buzzing Bangna-Trad motorway, redirecting their focus onto the coffee.

Through this concept, the boundaries between the three practices; architecture, interior, and landscape are blurred - the realism of the exterior and interior are connected. These "blurred" spaces create "undefined areas" where Instagram-ability is naturally made to be less important than the visitor's "experience" of indulging in a high-quality cup of coffee.

The main buildings in the front are a result of maintaining continuity in the roofline with the existing building, which extends into three masses where the Speed Bar and the indoor seats are located. Green areas of the landscape infiltrate the gaps between these masses and flow into the interior with the use of reflective glass mosaics on the ceiling.

The front part of the existing building is renovated as a part of the Slow Bar and service zone The restrooms located at the back of these buildings, are designed as independent pods, surrounded by the landscape for added privacy.

All buildings are designed to be simple and functional, this simplicity continues into the design of the interior, where the main focus still revolves around the coffee-drinking experience. The counters wrapped around the perimeter of the room to direct customers' attention to the coffee instead of having a conversation. The counters have uneven contoured surfaces, which causes the customer to concentrate on the placement of their coffee and the drinking experience.

The counter also doubled as "social distancing" while referring to the northern mountain range where the coffee beans are grown. Other elements such as signage are also designed based on this concept of "concentration", whereas most typical signage wants to grab attention, the signage at NANA Coffee Roasters is designed to contain hidden details which are only revealed when being focused on.

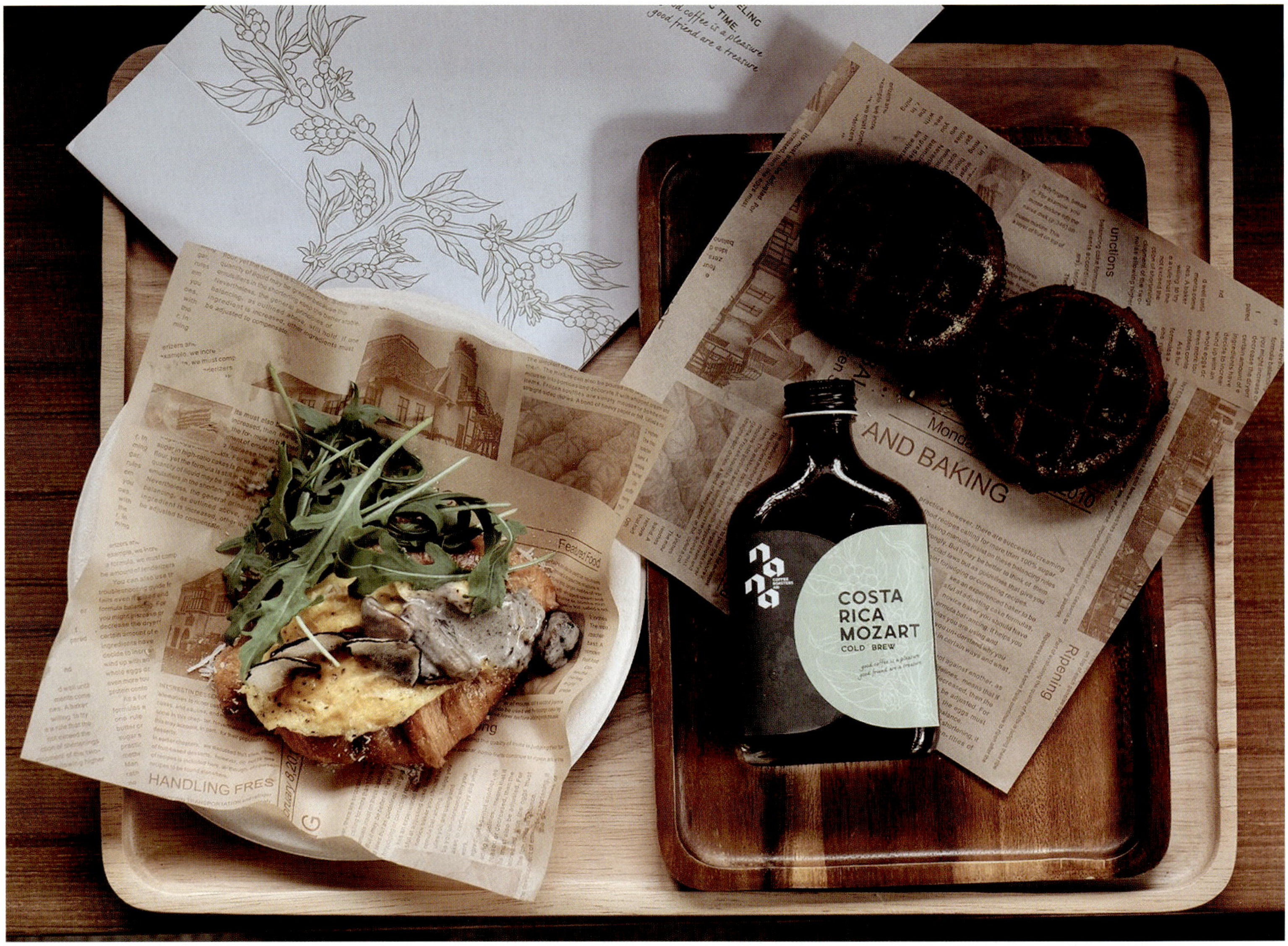
COSTA RICA MOZART
COLD BREW
AND BAKING
Ripening

WAKE UP
NITRO TEA
CHOCOLATE FLAVOR

esmeralda
jamaica
colombia
TASTING NOTES : TEA ROSE, PEACH TEA, STRAWBERRY MILK, MOLASSES, MILK CHOC, AYATAKA, APRICOT
ORIGIN : COLOMBIA
PROCESS : ANAEROBIC FERMENTATION +1,800 M.A.S.L
LIGHT ROAST

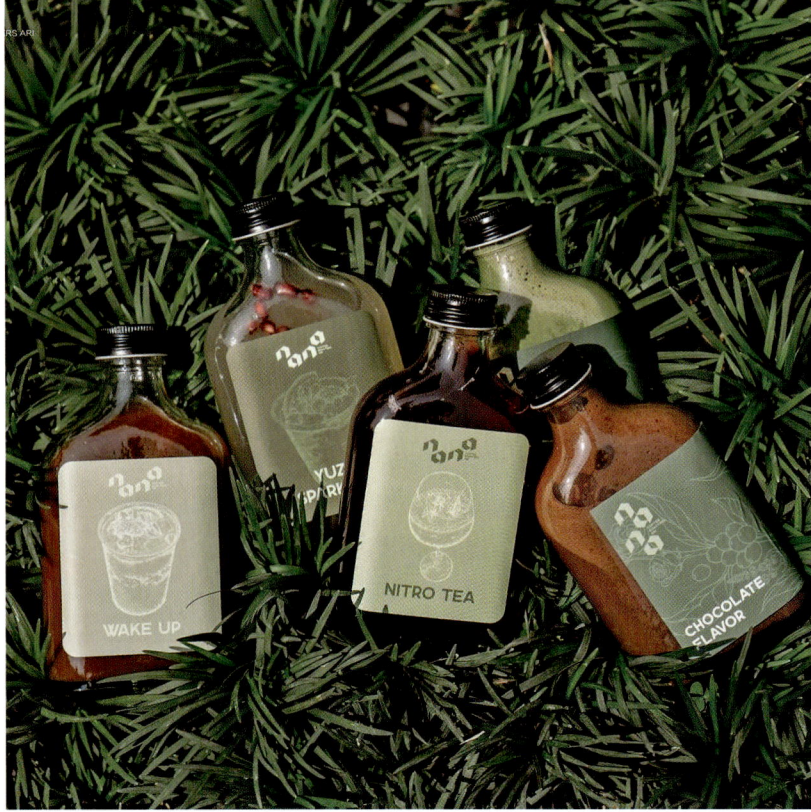

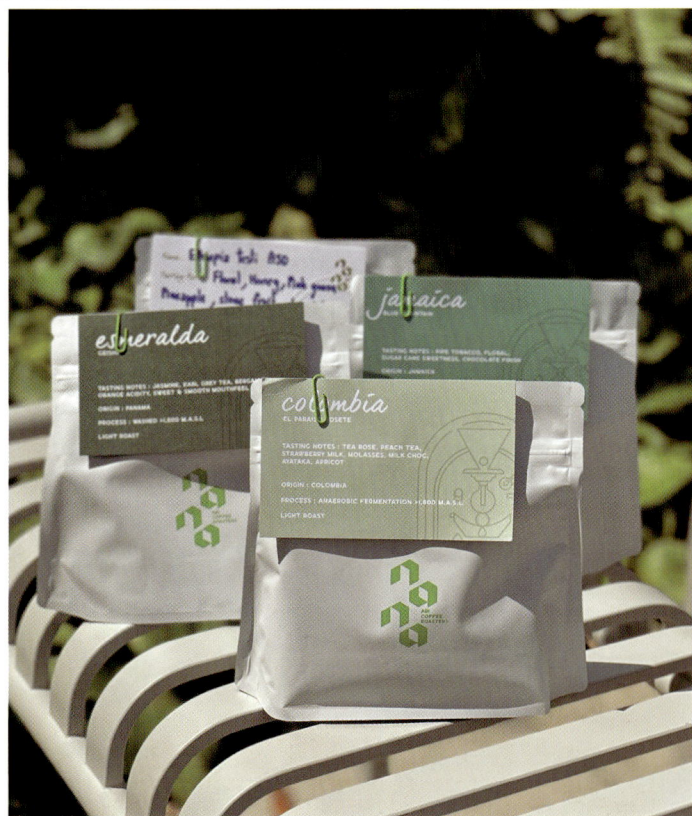

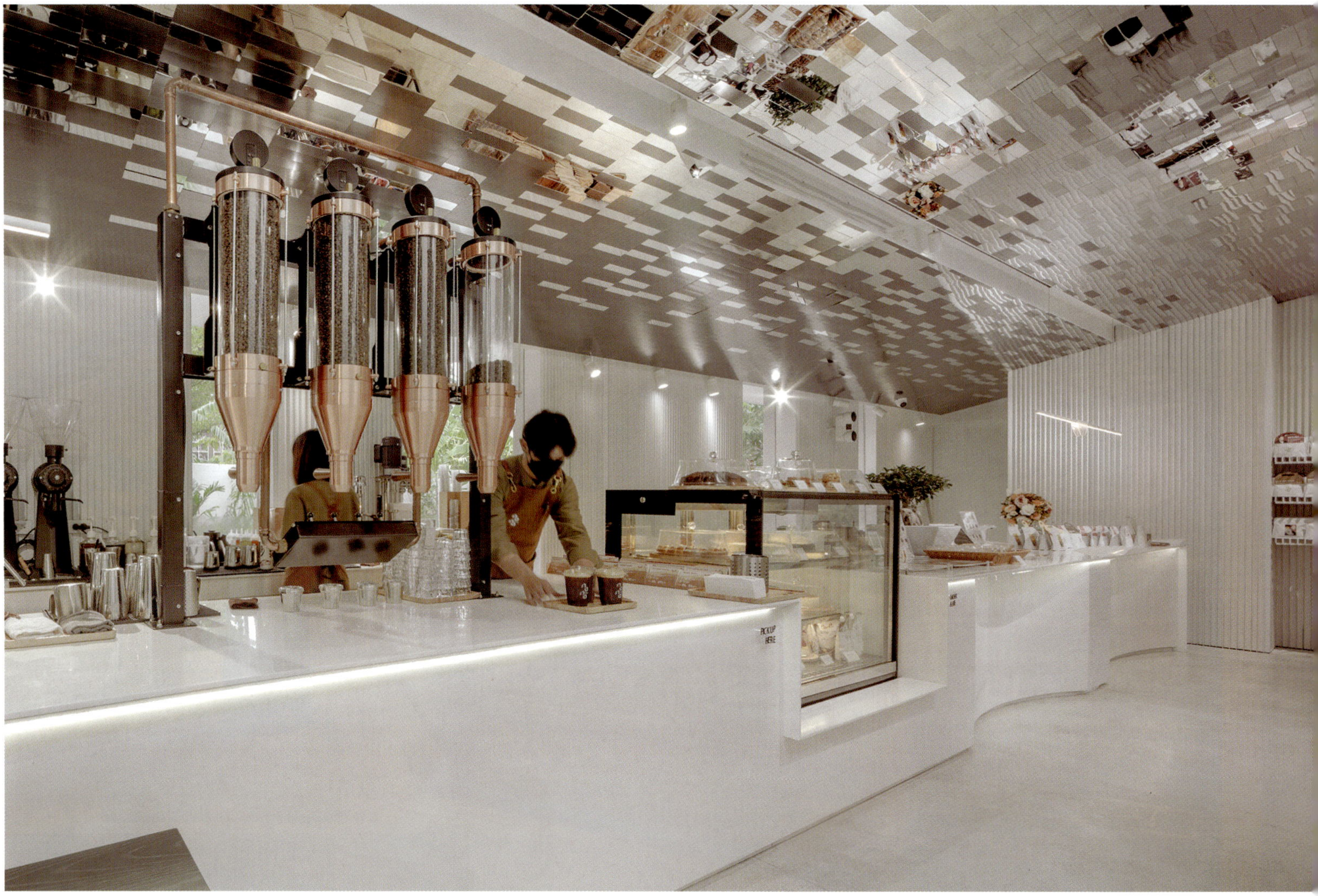

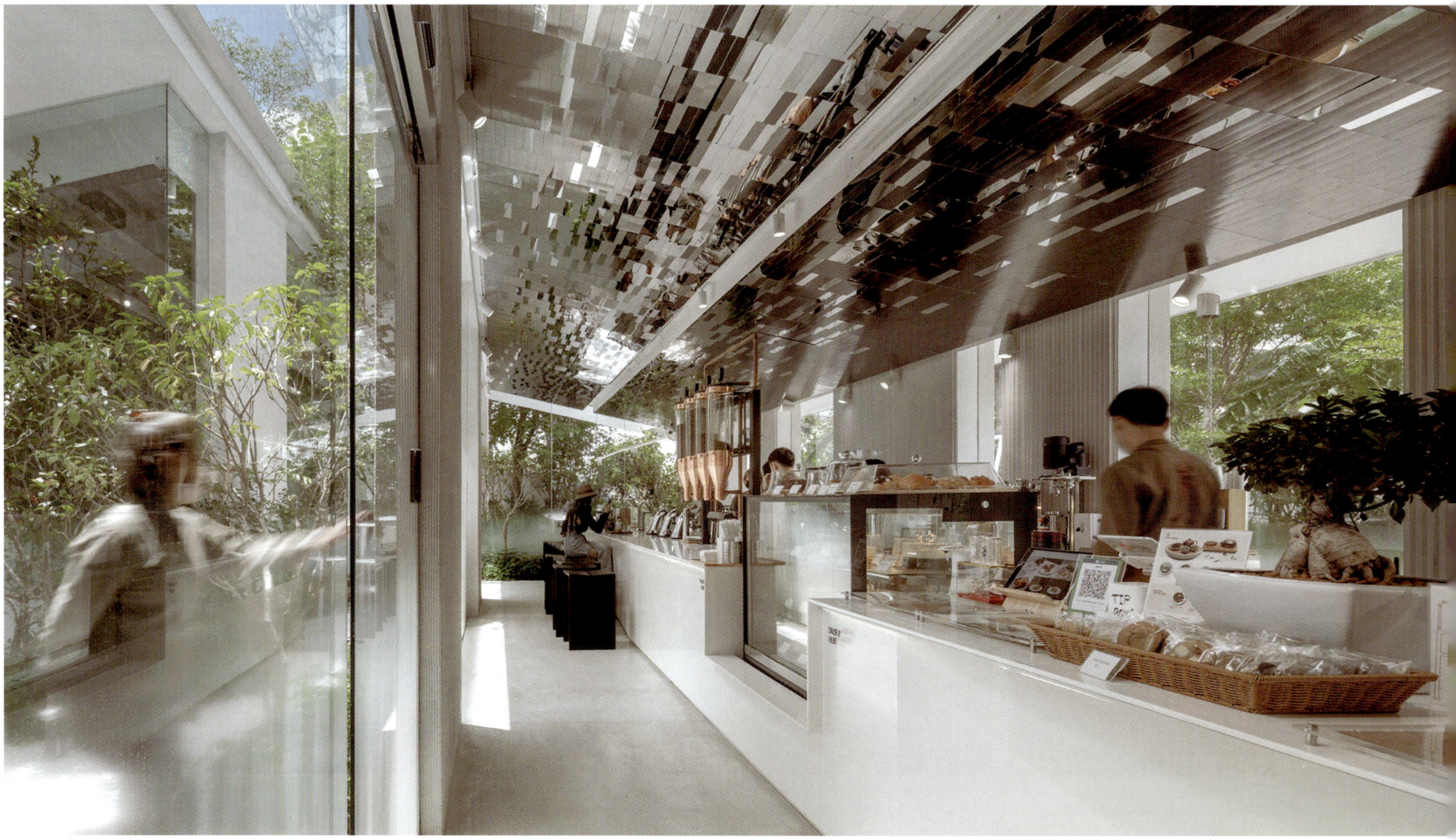

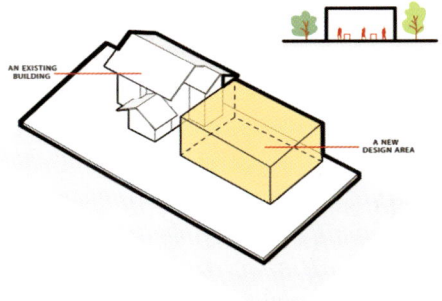

ONE MASSIVE SPACE

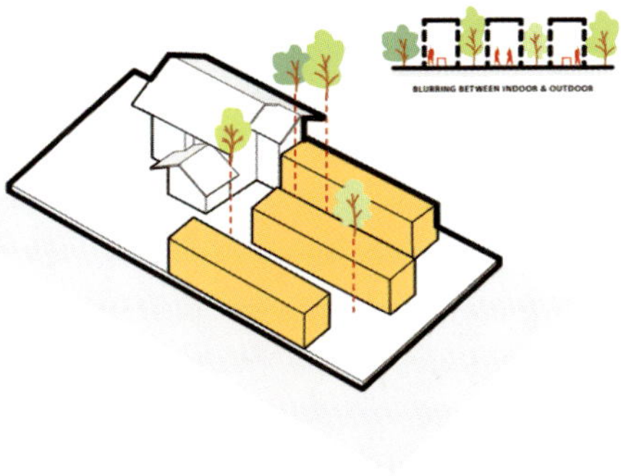

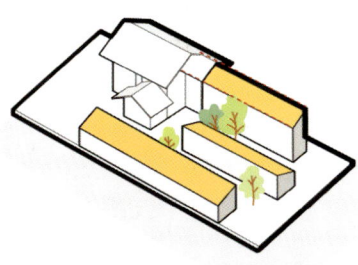

CREATING MORE PRIVACY BY DIVIDING DOWN THE SPACE AND ADDING GREENERY

ADAPTING THE LANGUAGE OF AN EXISTING ROOF TO THE NEW BUILDING

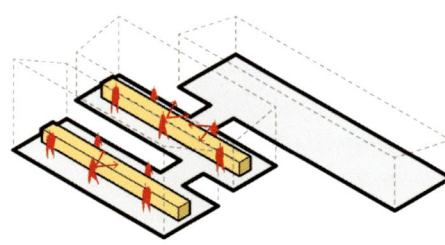

CENTERED ARRANGE PLANNING PACKED TRAFFIC, DISTURBING ACTIVITIES, AND DISTRACTIONS

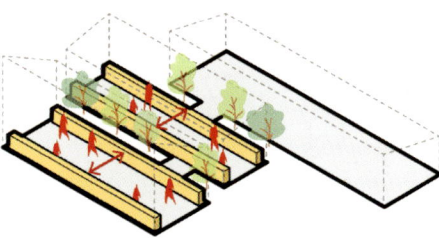

SET ASIDE FURNITURE TO THE EDGES CONTROLLED CIRCULATION, PRAISING VIEWS, AND CONCENTRATION

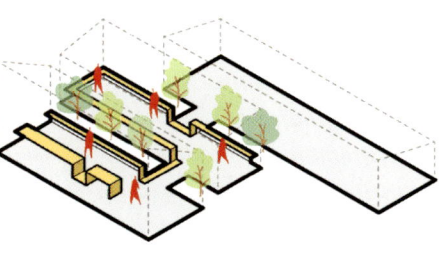

DRAG OUT SOME PARTS OF FURNITURE TO CREATE CONTINUOUS CIRCULATION

⑦ Sizing
⑧ Hand Sorting
Moka Pot
Espresso
①Cleaning
COFFEE PROCESSING
②Pulping
③Fermentation
④Cleaning
Dry Process
Honey Process

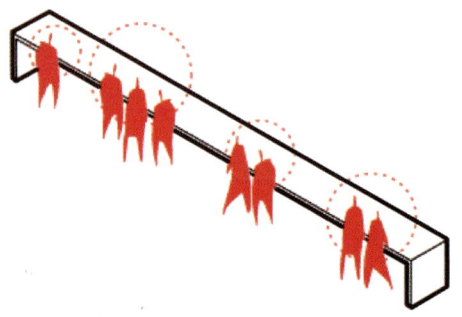
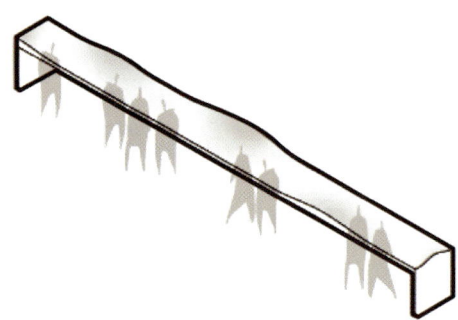

THE CONTOURED COUNTER TOP FOR THE USERS TO CONCENTRATE ON THE COFFEE WHICH ALSO SERVES AS A WAY TO "SOCIAL DISTANCING", ALSO REFERING TO THE MOUNTAINS RANGE, THE ORIGIN OF THE COFFEE.

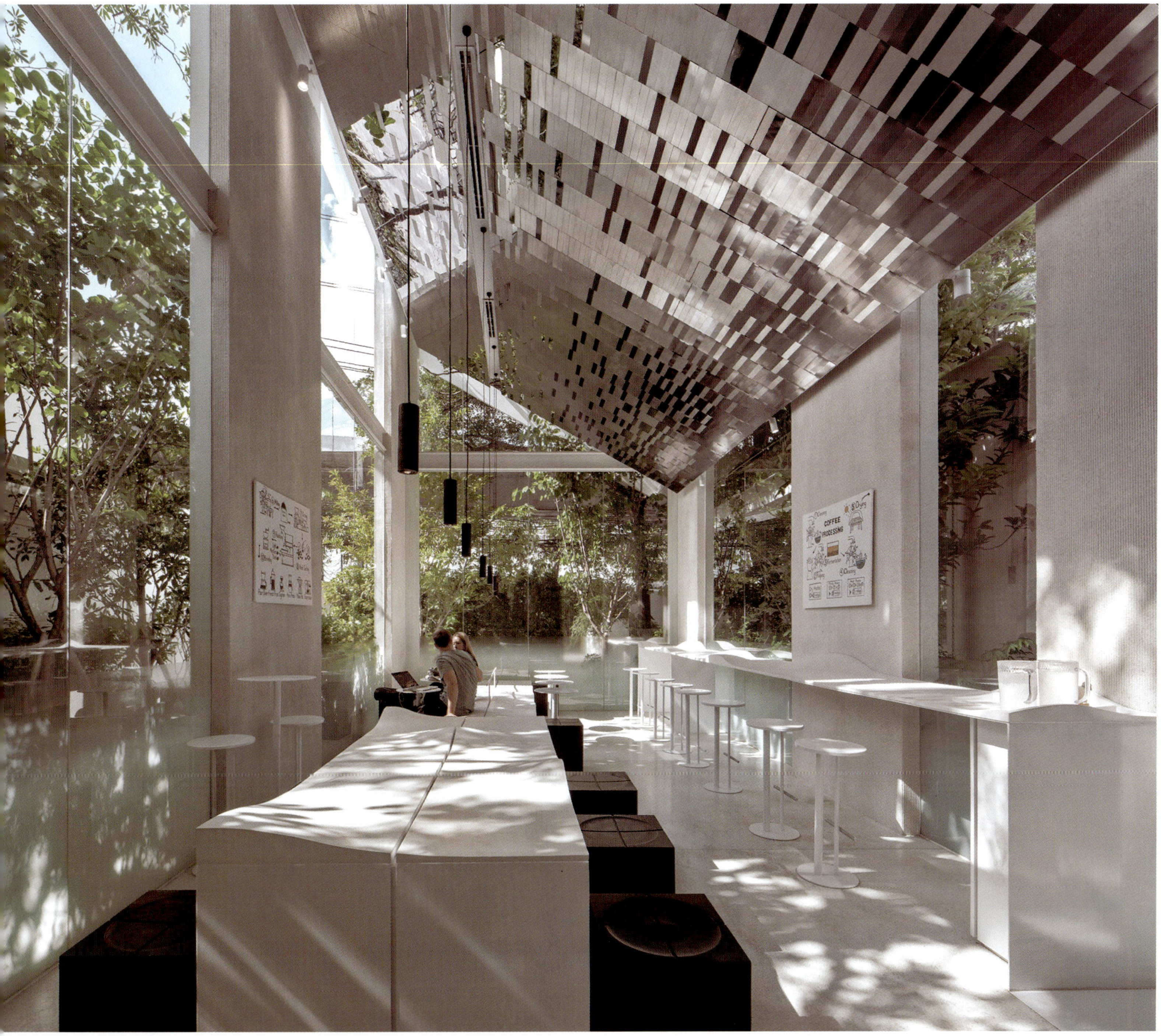

PICK UP
HERE
COSTA
RICA
MOZART

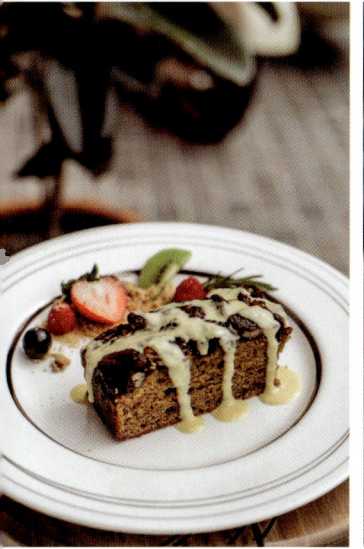

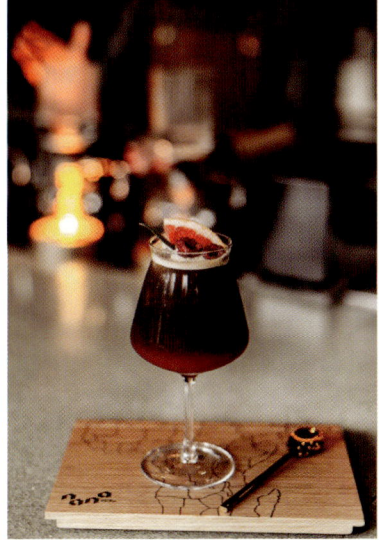

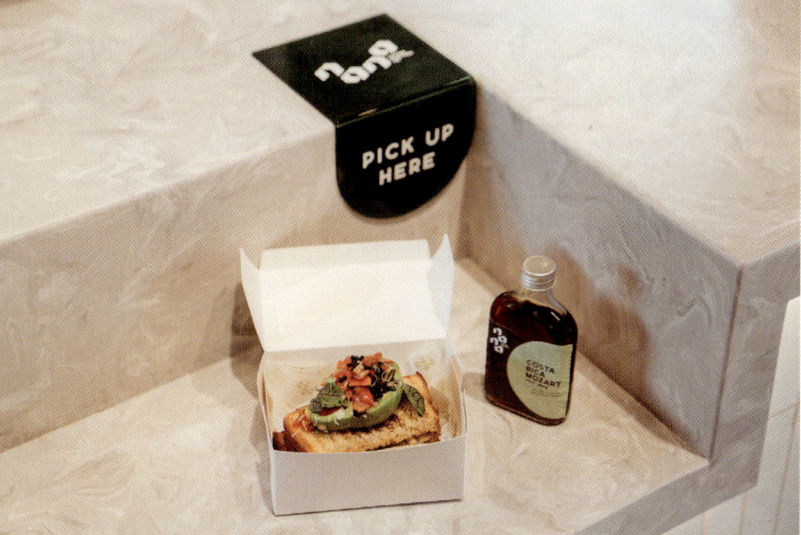

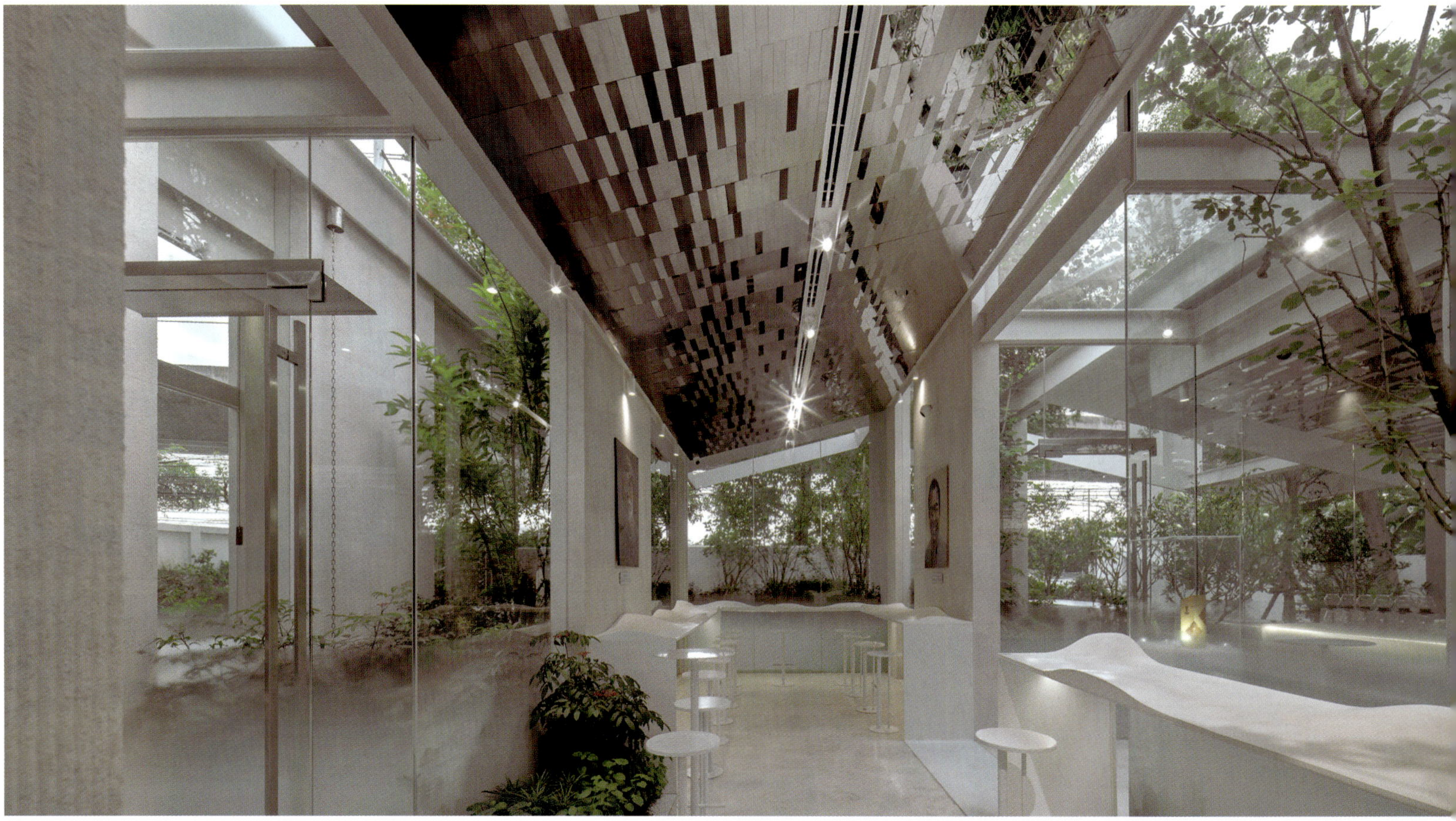

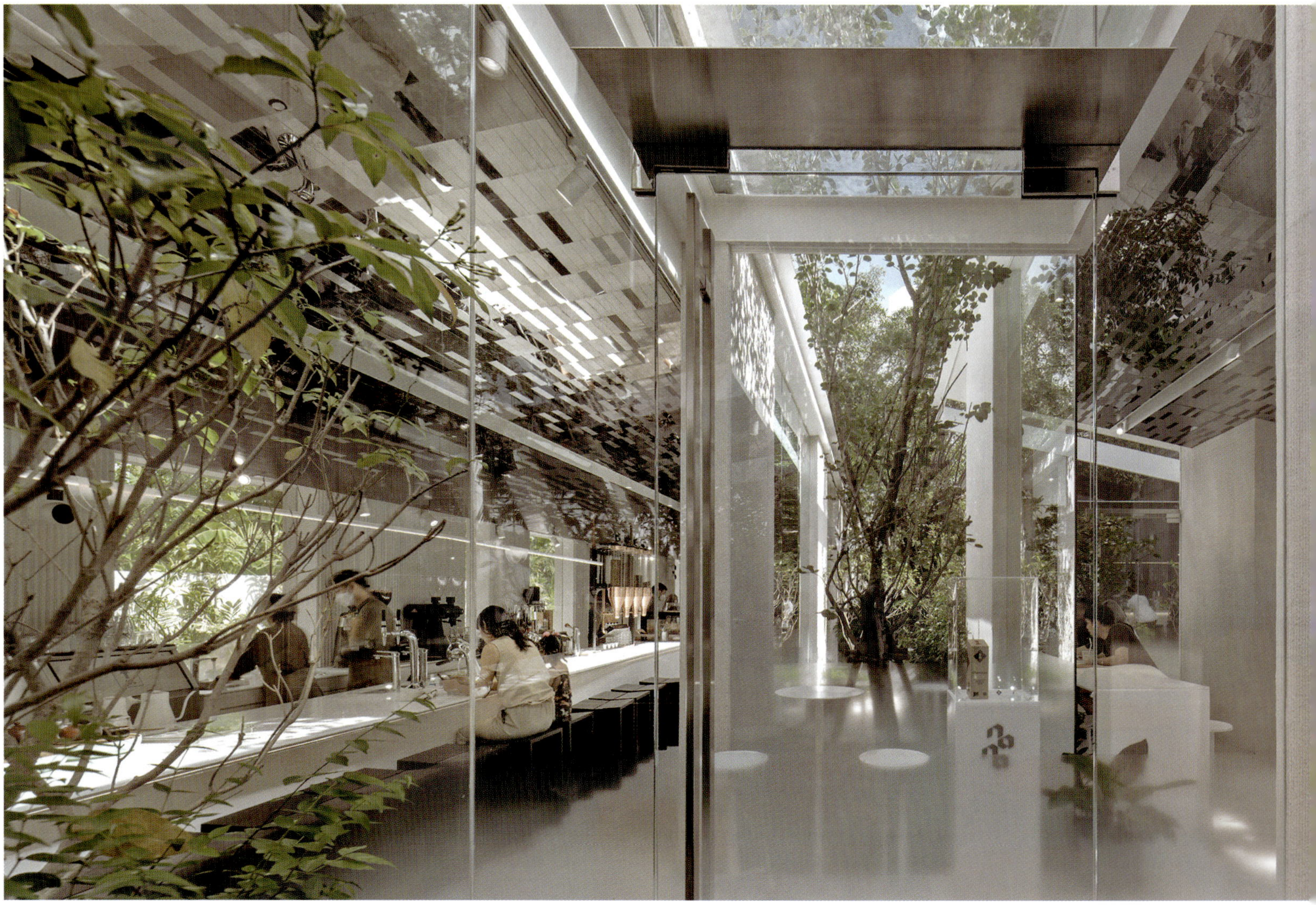

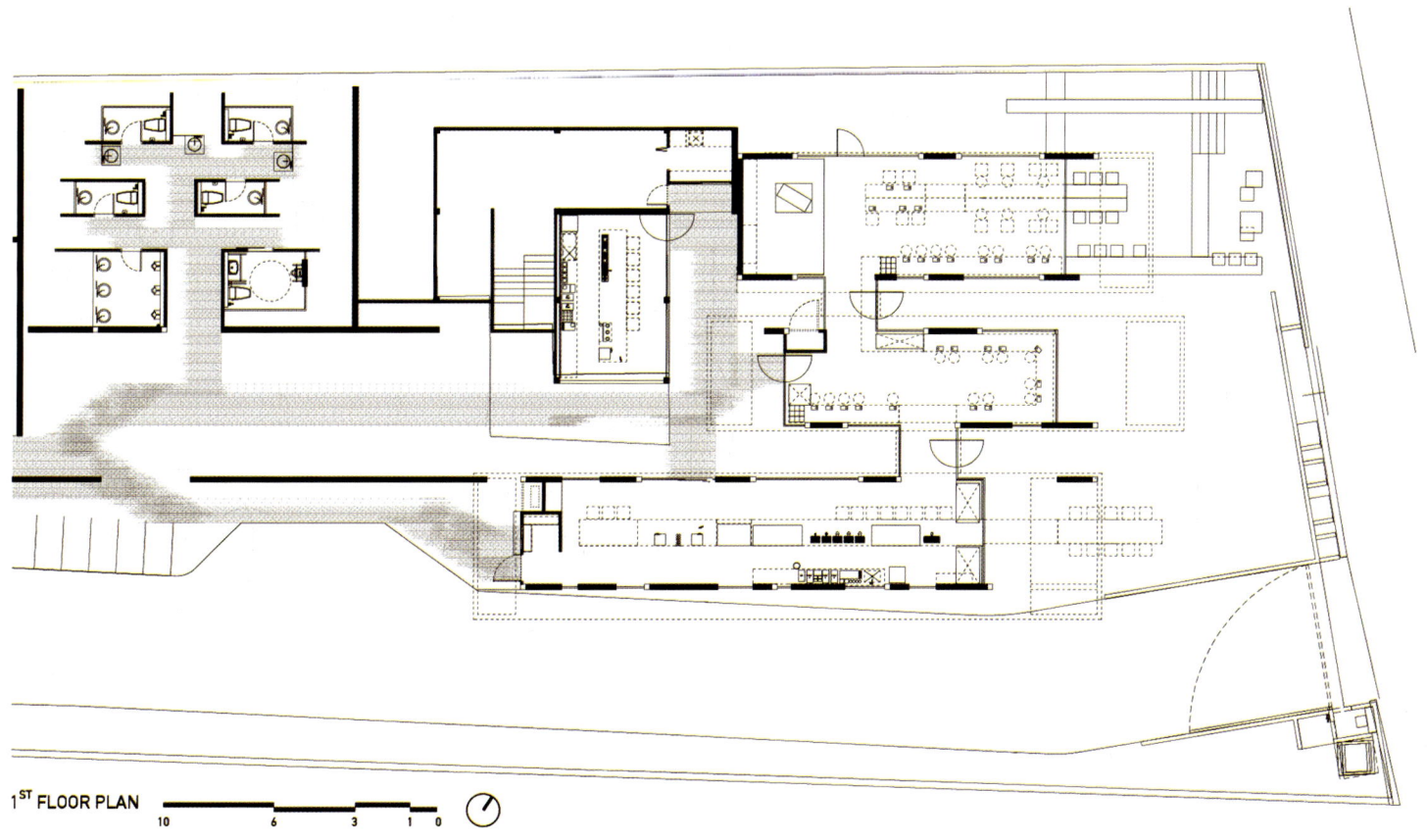

1ST FLOOR PLAN

Site Plan

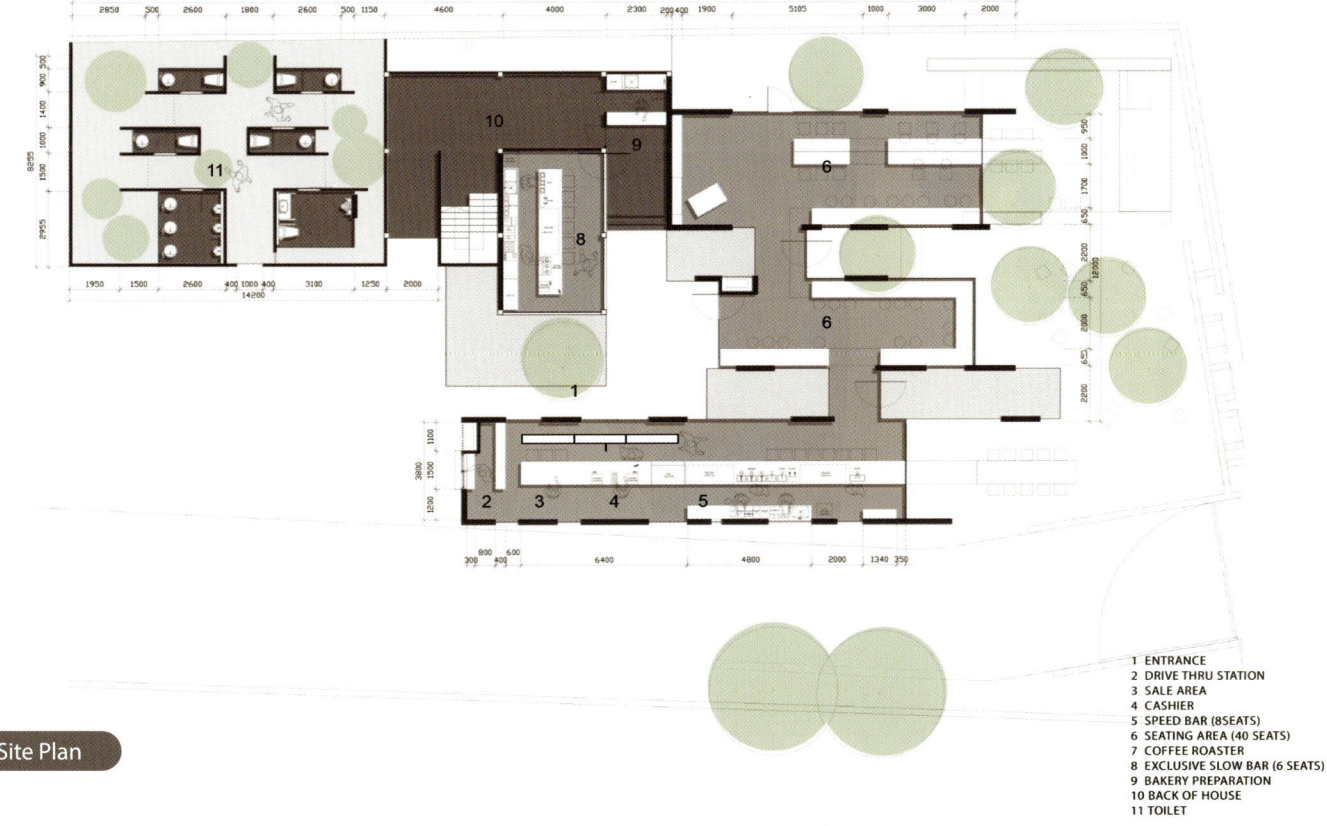

1 ENTRANCE
2 DRIVE THRU STATION
3 SALE AREA
4 CASHIER
5 SPEED BAR (8 SEATS)
6 SEATING AREA (40 SEATS)
7 COFFEE ROASTER
8 EXCLUSIVE SLOW BAR (6 SEATS)
9 BAKERY PREPARATION
10 BACK OF HOUSE
11 TOILET

Primo Cafébar
Tübingen, Germany

DIA - Dittel Architekten?

An acronym for Integrating Design Into Nature, IDIN Architects was founded in 2004. We perceive 'nature' in two ways. Firstly, nature can be defined as the ecology around us. Secondly, it can also refer to different mannerisms and personalities. The design philosophy and attention of IDIN are to merge this sense of surroundings, the 'natures', to the architectural aesthetic. This merge is done through a process of analyzing and prioritizing the different needs and requirements of each project.

In addition to being an acronym, the Thai word "idin" is used to describe the natural phenomenon when soil releases a beautiful scent after rainfall. This symbolizes Thailand's tropical climate, something all IDIN designs aim to respond to. Our emphasis is therefore placed not only on aesthetics, but also on being practical, in order to suit this temperate environment.

Design: DIA - Dittel Architekten **Homepage:** https://di-a.de **Image:** DIA - Dittel Architekten
Location: Tübingen, Germany **Team:** Frank Dittel, Biance Unger, Lennart Schutz, Jana Schmalohr....

The Primo Cafe Bar stands for high-quality coffee, an Italian lifestyle and a sustainable mind-set. In the spirit of this brand philosophy, an authentic interior design concept with natural materials and dedication to detail was created. The café was realized at the Zinser fashion store in Tübingen. Located in the pedestrian zone, it not only attracts people from the outside but also creates value for the store's retail customers. The continuous glass façade reveals a warm atmosphere to passers-by which - in a symbiosis of light and material - makes them want to enjoy a coffee. The concept places the service counter at a central location, because this is where the processes of the café's operations come together. This core function makes it a formative element that integrates all materials. Inspired by the traditional coffee house culture, ceramic tiles in the chessboard pattern highlight the more than 10-metre-long l-shaped counter. The counter is clad in whitened oak slats, creating a natural colour spectrum with a lively effect.

The shape and material of the carcass are finished off with a massive wooden top and an inserted glass display case. The suspended black steel shelves in the background have the charm of untreated materials and provide an effective contrast to the soft colours of the organically shaped ceramic tiles.

The cafe area is divided by means of vertically arranged wooden slats which provide a reference to the structure of the counter and a view of the sales area. The visitor can choose to sit on an raised platform with a view of the outside, in comfortable leather chairs in the lounge corner or in the interior of the room for a social get-together. A thematically adapted ceiling element with unusual catenary lights supports the cosy atmosphere. Golden accents in the lights take up the packaging design of the products and represent the exquisite coffee culture. The robust appearance of the wood-branded logo stands for craftsmanship and nature.

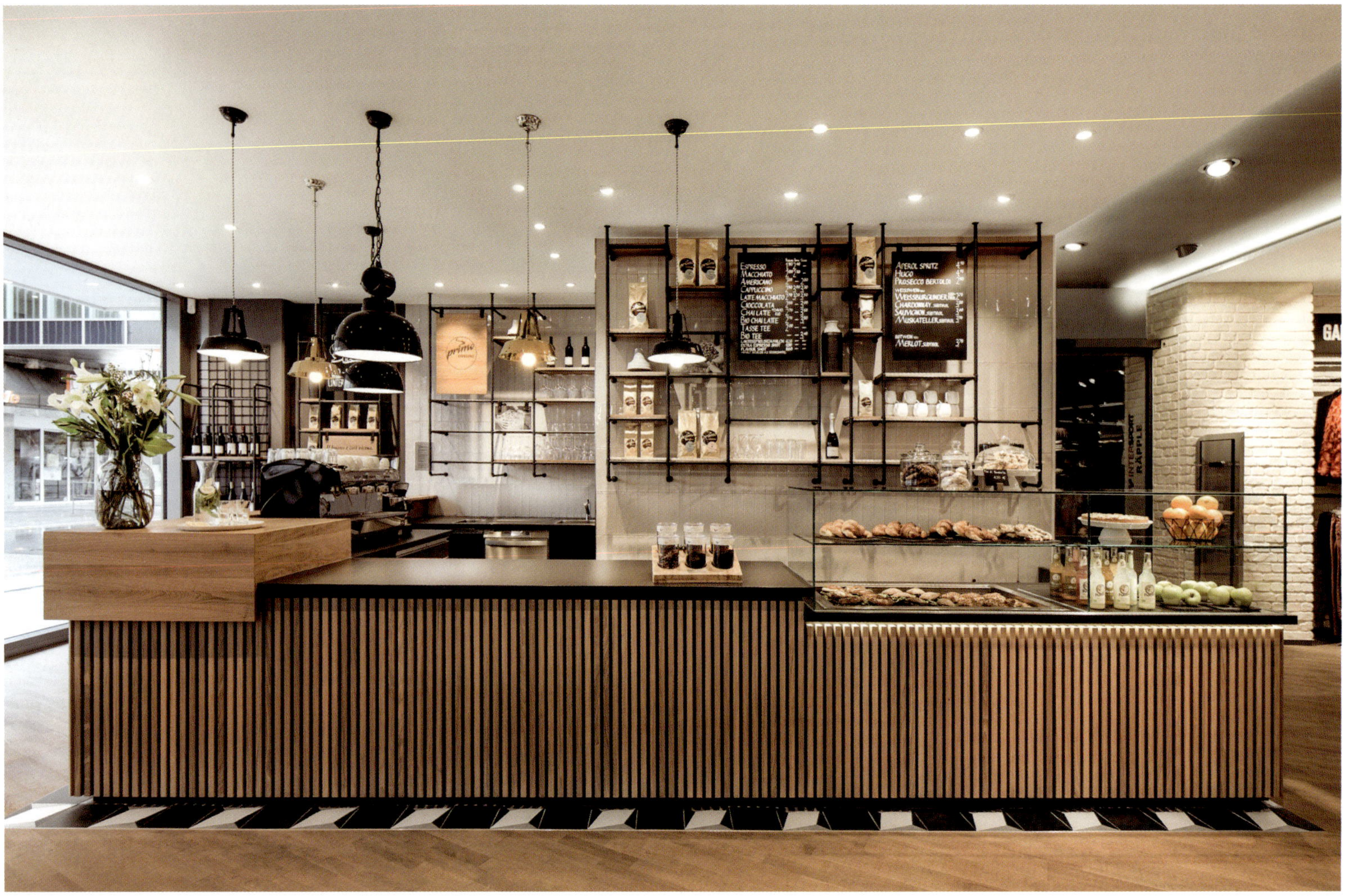

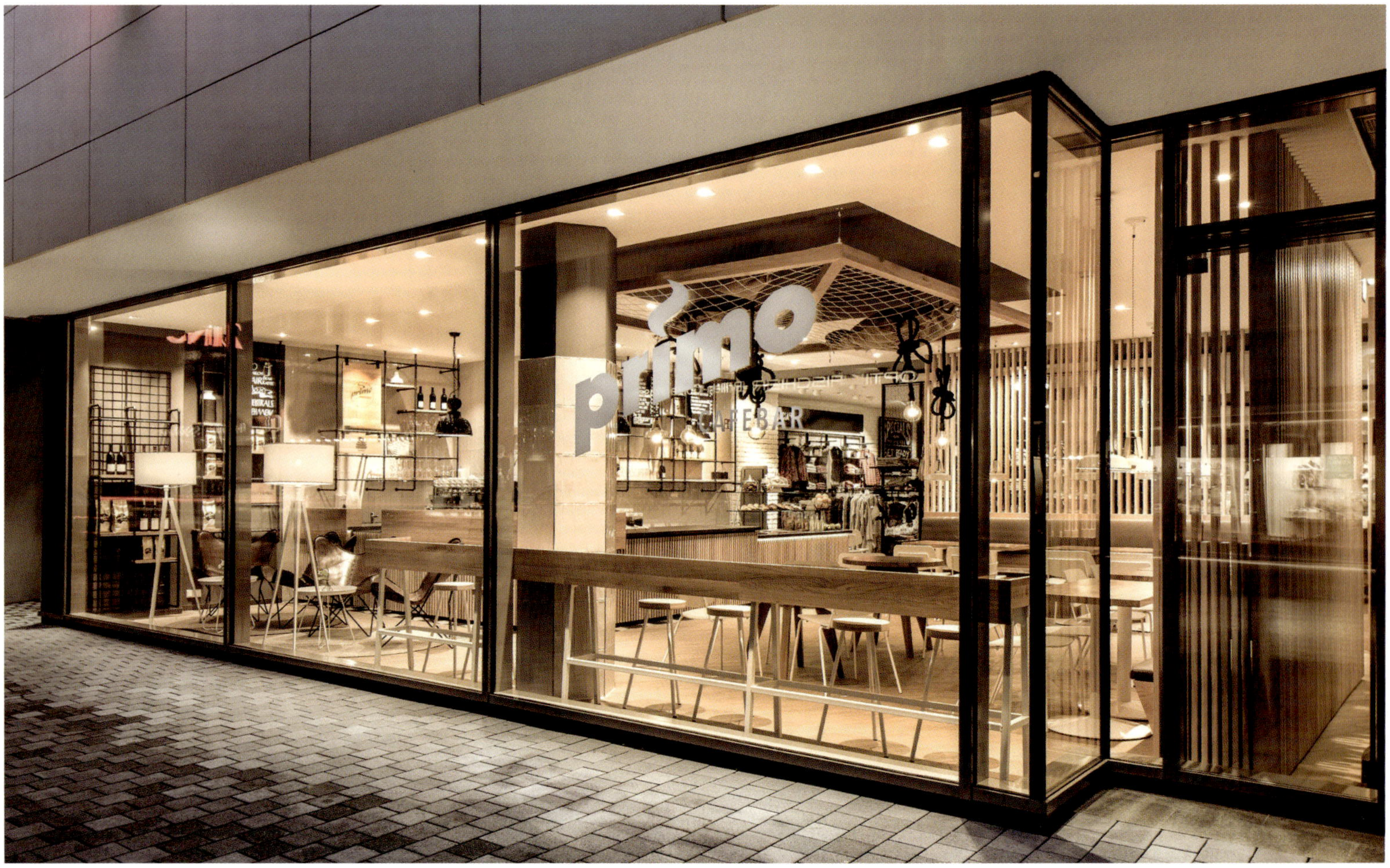

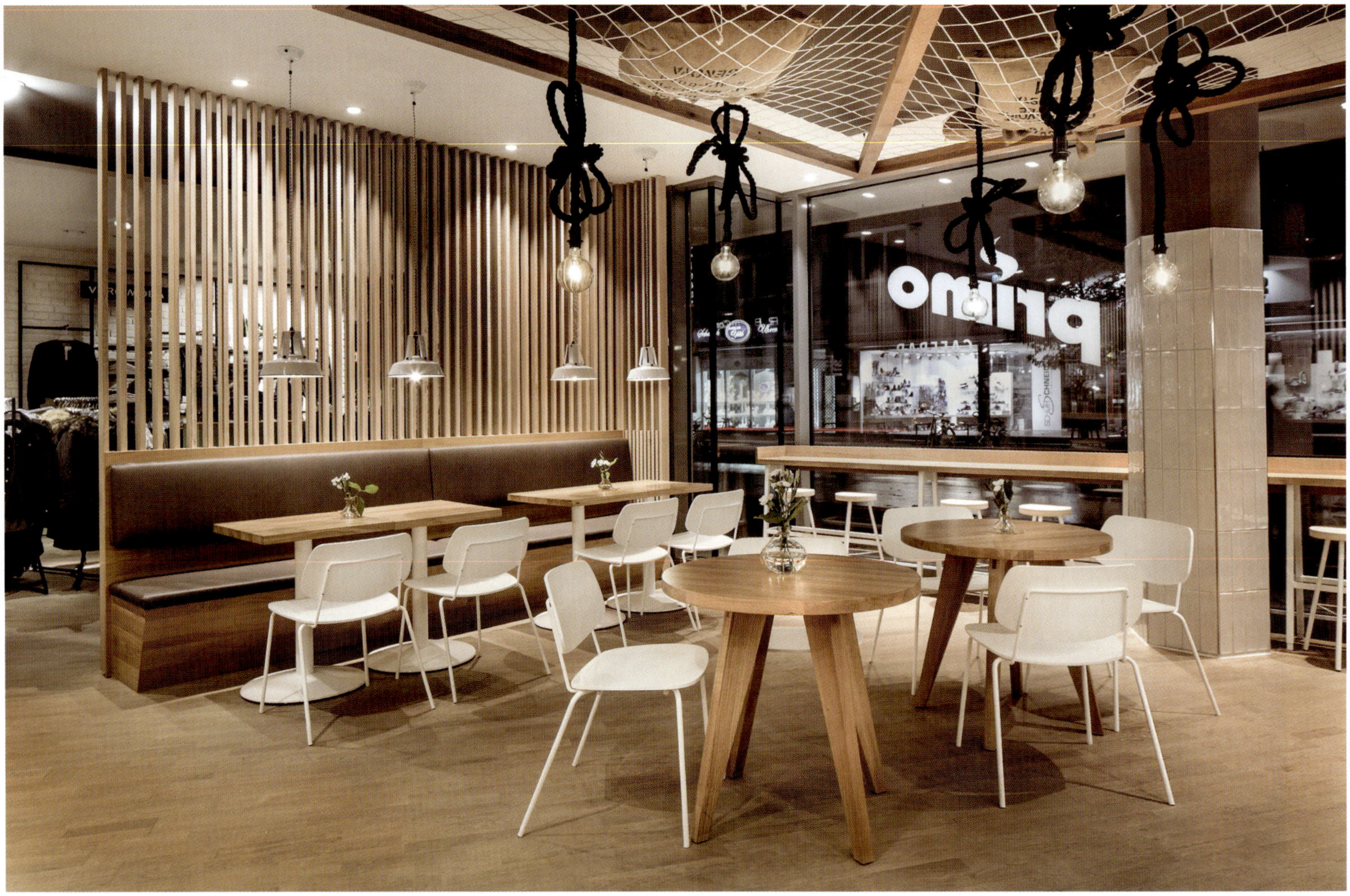

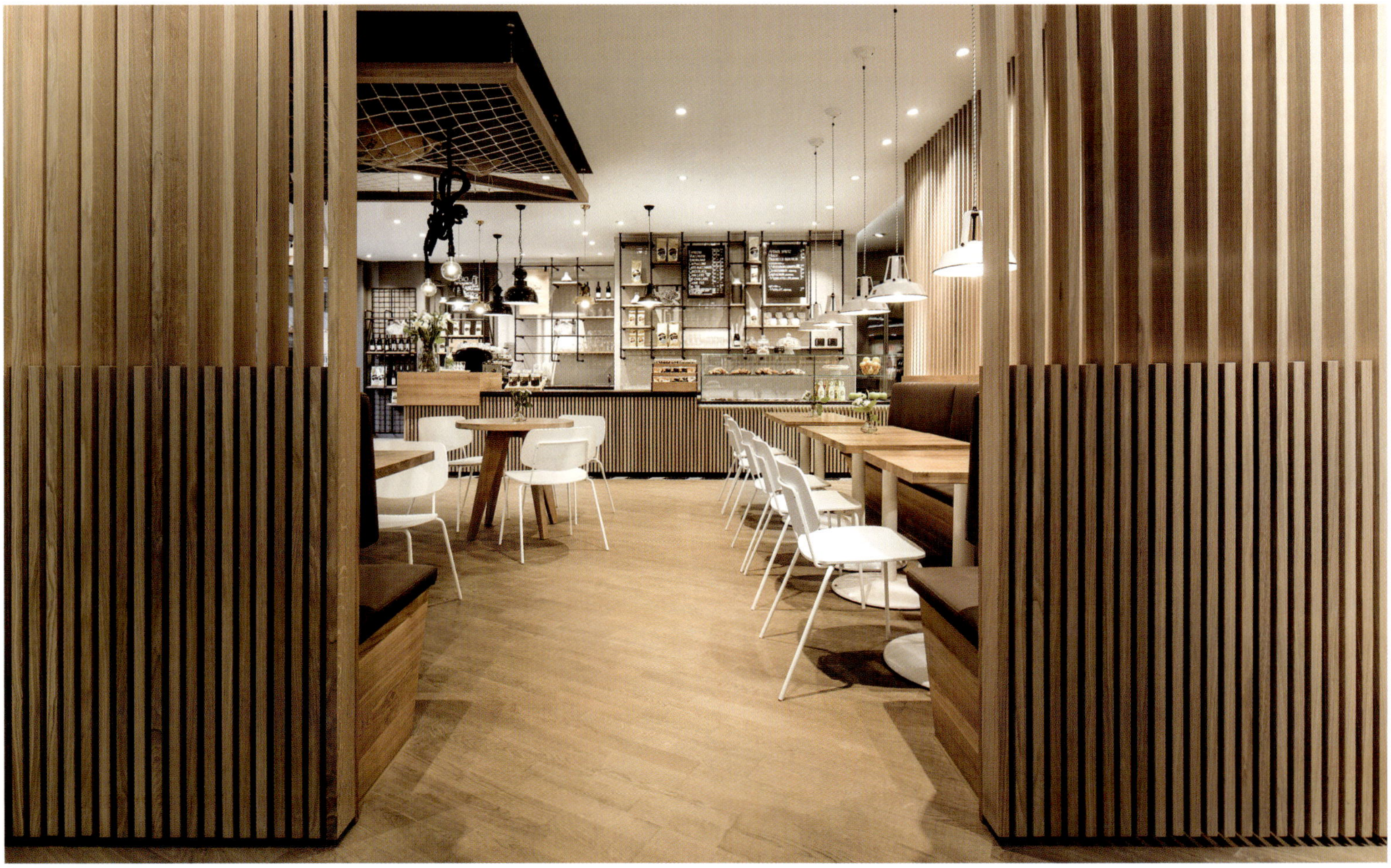

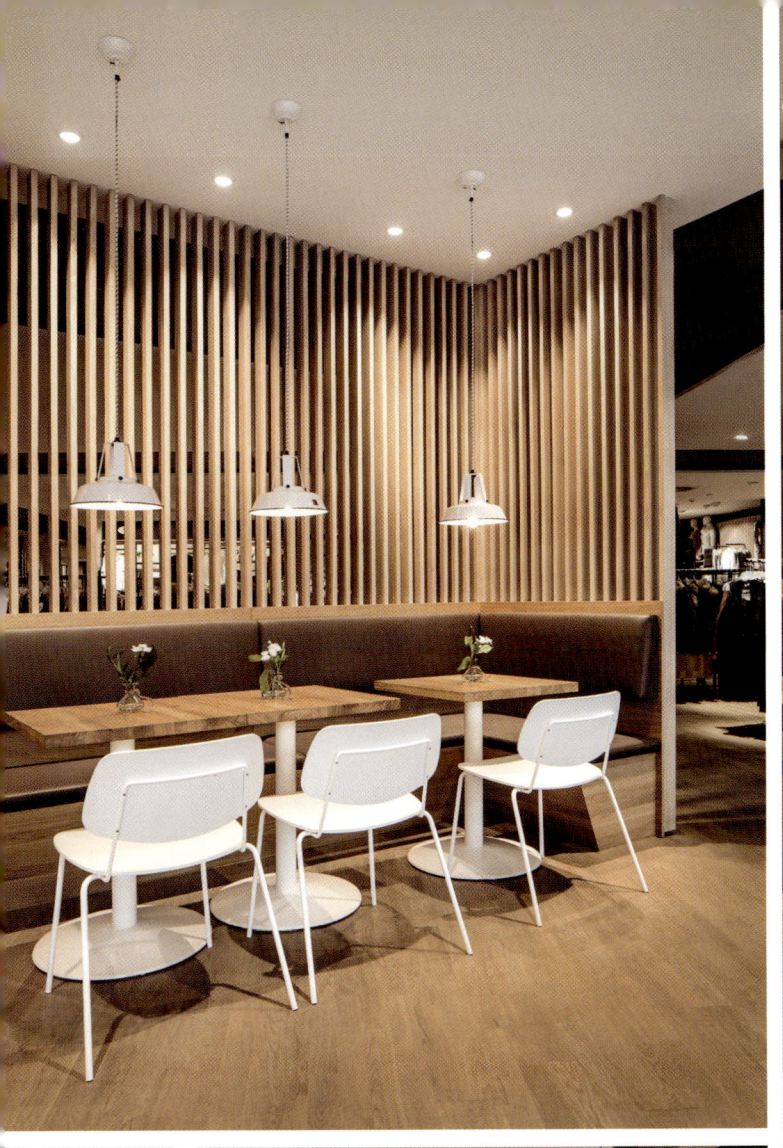

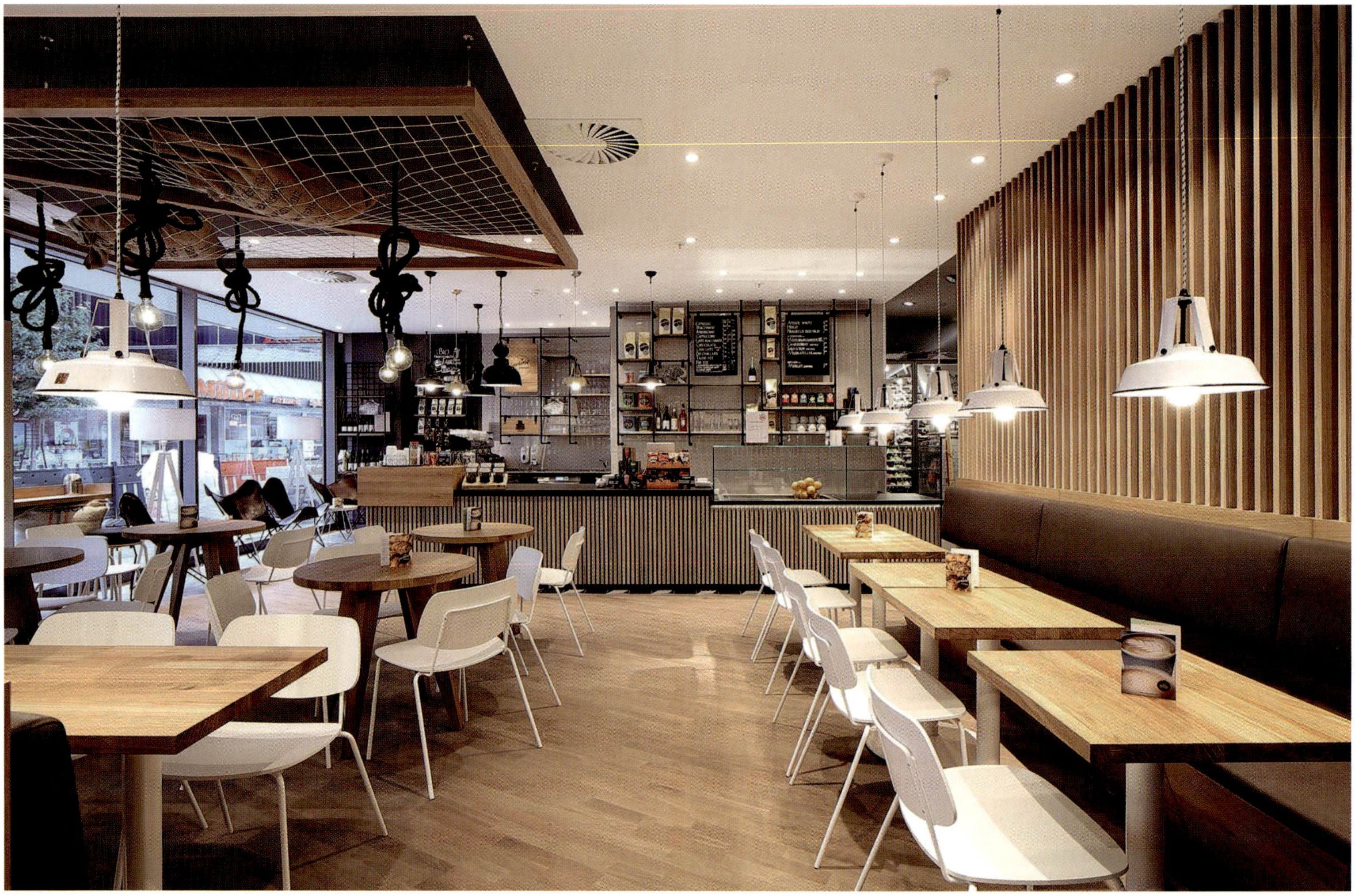

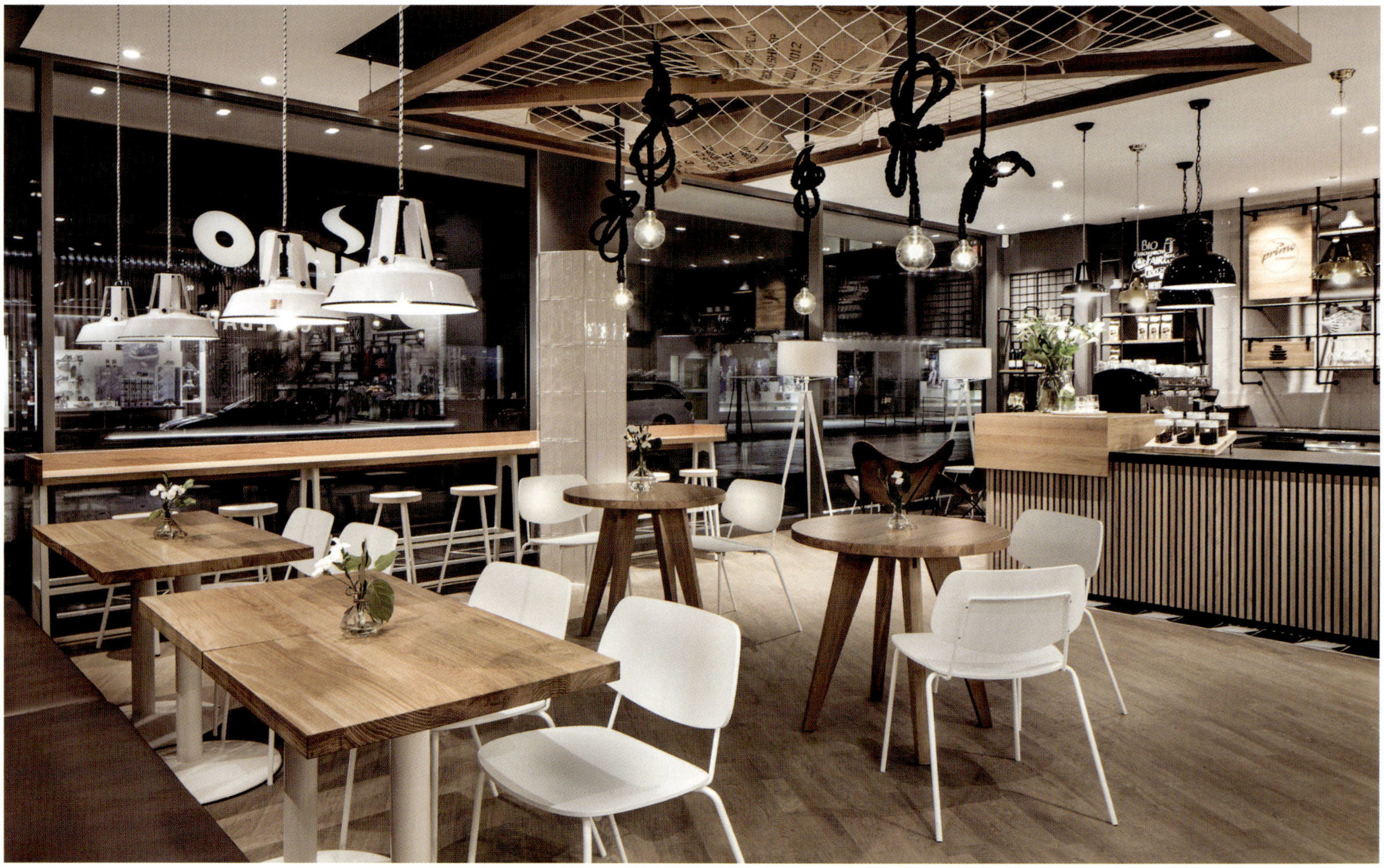

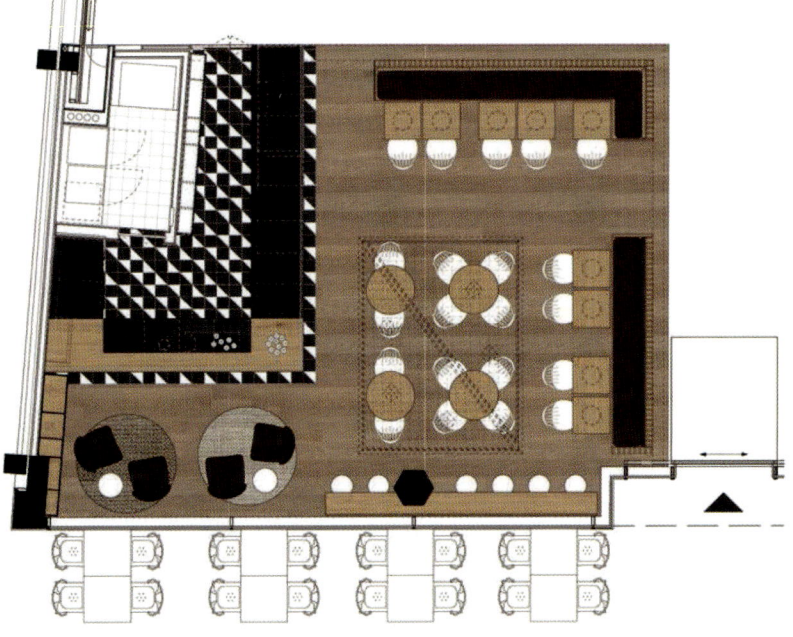

Floor Plan

Cupping Room
Palace Museum, Hong Kong

M.r. Studio?

M.R. Studio Ltd. (M.R.S) is a design studio known for its specialization in hospitality, commercial, private residential, and office projects. Founded by award-winning designer Myron Kwan, he creates multi-faceted couture interiors, advising on every aspect and detail of a project through his expertise in space planning, conceptual and design development, and instilled with a heightened sense of narrative. Our style is, we believe, enthralling, elegant, and timeless. In which, the work of M.R. Studio employs a range of ideas, lines, materials, objects, and references – from unexpected to ever-evolving. We believe every element of our handcrafts environments enable our very appreciative clientele to share our obsession with detail, craft and refinement.

Design: M.r. Studio **Homepage:** https://mrstudio.hk **Photography:** M.r. Studio
Location: Palace Museum, Hong Kong **Area:** 260m2

Cupping Room was founded in 2011 in Stanley Plaza, Stanley, Hong Kong Island. From the beginning, our focus was to bring carefully roasted and prepared coffee to Hong Kongers in a comfortable environment. In 2013, we relocated from Stanley Plaza to the bustling district of Sheung Wan, expanding in the process to include Australian-inspired café fare. In 2014, we added our first location in Wan Chai and in 2015 our Central store soon followed, which included an on-site bakery. Finally, we welcomed our Roastery and coffee bar at Po Hing Fong in 2016 and have been roasting coffee since then on a Probat UG15 Retro roaster.

We welcomed our first shopping mall locations in Lee Garden One, Causeway Bay and Harbour City, Tsim Sha Tsui. Our Lee Gardens shop offers location-exclusive Nordic-inspired open face sandwiches, while our Harbour City cafe features panini sandwiches grilled to order. Both serve our newest house blend, "B-side" (read more about B-side here) in all espresso-bar beverages. Approaching our 10th anniversary, we're celebrating by refreshing our look with a new logo and style! It's an exciting time for us as we plan for the next 10 years...

The interior design concept is to serve both aesthetic and cultural purposes. The designer deftly blends elements of traditional Chinese architecture, ancient Chinese philosophy and culture with a distinct modern classy vibe into the design of the space. The elegant decoration style fits the temperament of the Forbidden City. The four treasures of the study are displayed under the glass table. The walls are also spliced with blue and black tiles. There are many outdoor seats, which are suitable for enjoying the sea breeze in good weather.

The interior design concept is to serve both aesthetic and cultural purposes. The designer deftly blends elements of traditional Chinese architecture, ancient Chinese philosophy and culture with a distinct modern classy vibe into the design of the space. The elegant decoration style fits the temperament of the Forbidden City. The four treasures of the study are displayed under the glass table. The walls are also spliced with blue and black tiles. There are many outdoor seats, which are suitable for enjoying the sea breeze in good weather.

September Coffee Shop
Vinh Hoi street, Australia

September
CAFÉ & CAKE

Red5studio + Ben Decor?

Our definition of a beautiful design emphasizes on the relationship between human and space, activities and environment rather than meaningless decoration in order to deliver clients true experiences of happiness via their interactions with the spaces. Each project is not only a challenge or an opportunity, but also a story to tell. And we want to become a greatest storyteller

Design: Red5studio + Ben Decor **Homepage:** https://www.red5studio.vn **Photography:** Phú Đào
Location: Vinh Hoi street, Vietnam **Area:** 260 m²

September is the cafe brand that Red5 has accompanied from the first coffee shop. At the next coffee shop, we bring the story of "The Wind and the Nest" to continue the story "Autumn and Fall Leaves" from the previous store. With the current status of two adjacent houses, off-floor dams connected, we want to create a nest with many nooks and crannies throughout the two houses. With the advantage of width and height, the façade repeats the bird's nest's image with a different perspective and uses many surrounding steel systems.

The circles hanging in the façade will move as the wind blows like a symbolic image of birds on a branch. To create the familiar and characteristic feeling of September, we use gentle neutral tones such as white, beige, rose-orange, natural wood colors.

Besides, the circular image and the curve reminded through the glass holes of the entrance, stairs, on the wall or even just the tiny details on the furniture, the wind's image is led by curves in space, through the ceiling, walls, and floor, creating a light feeling like a breeze spreading into each corner. Regarding the layout of the space, the ground floor with the spirit of bringing the garden into the house creates an "indoor but outdoor" feeling suitable for guests to sit fast or sit around the bar looking at mixing. The 2nd and 3rd-floor areas are for those who sit longer or go in groups. If the rooftop is an aerial garden, the lamp section is like a branch supporting the nest, and this is also the highlight of the space for those who love to check in and is the smoking area. The entire furniture and space had inspired by bird's nests, curves, branches, birds on branches or simply a tiny dot circle, all creating a bird's nest in the heart of the city.

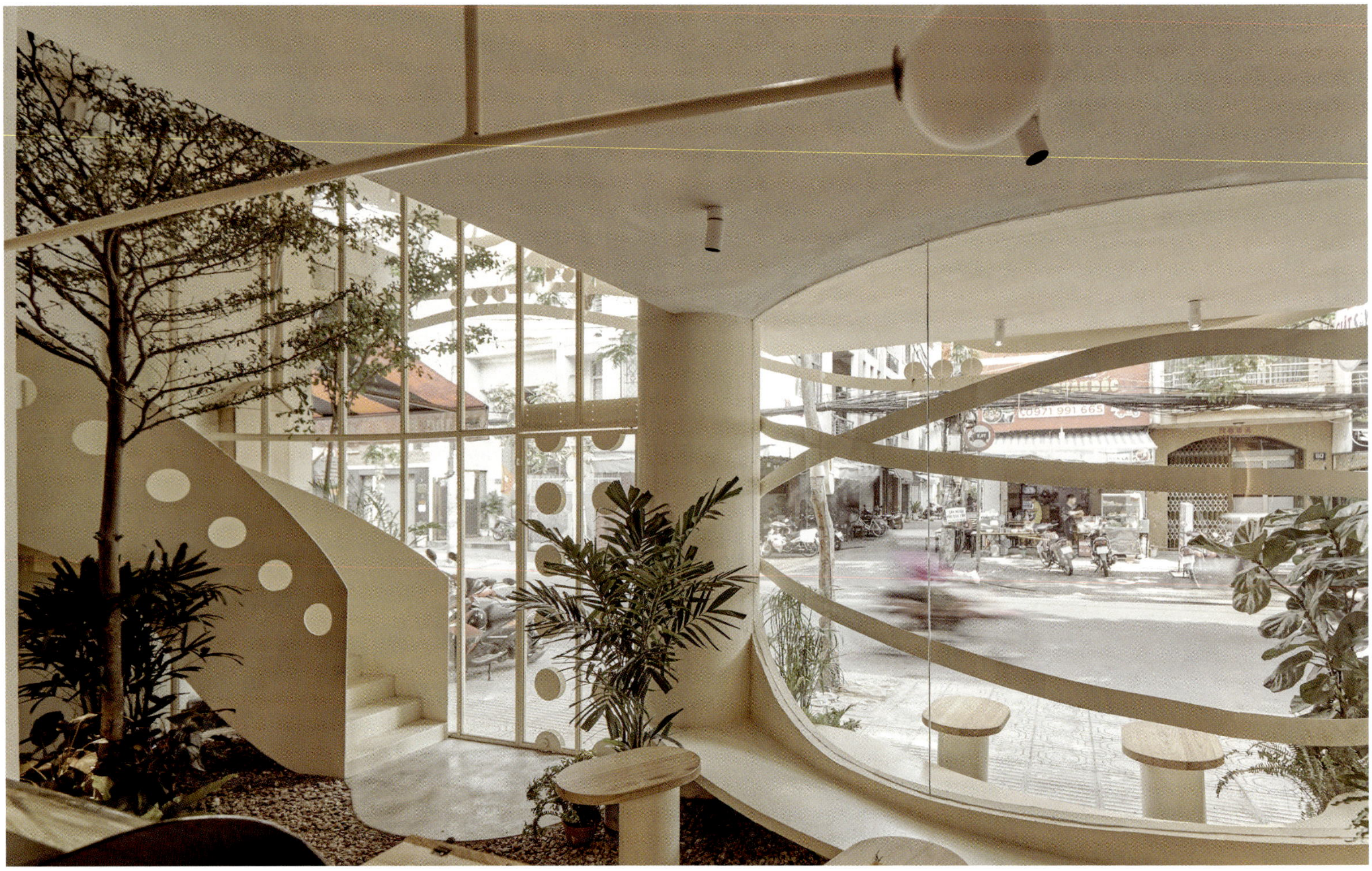

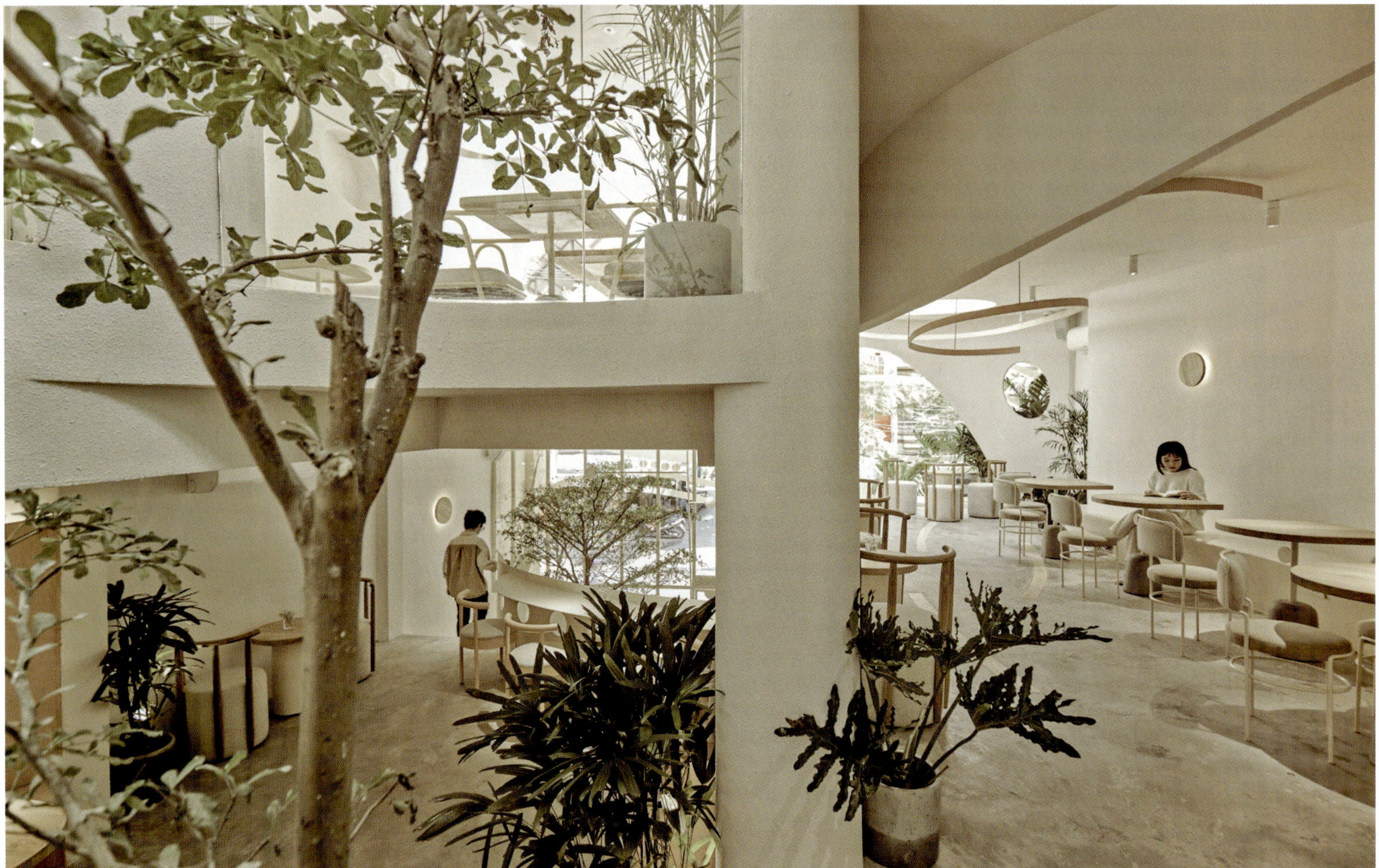

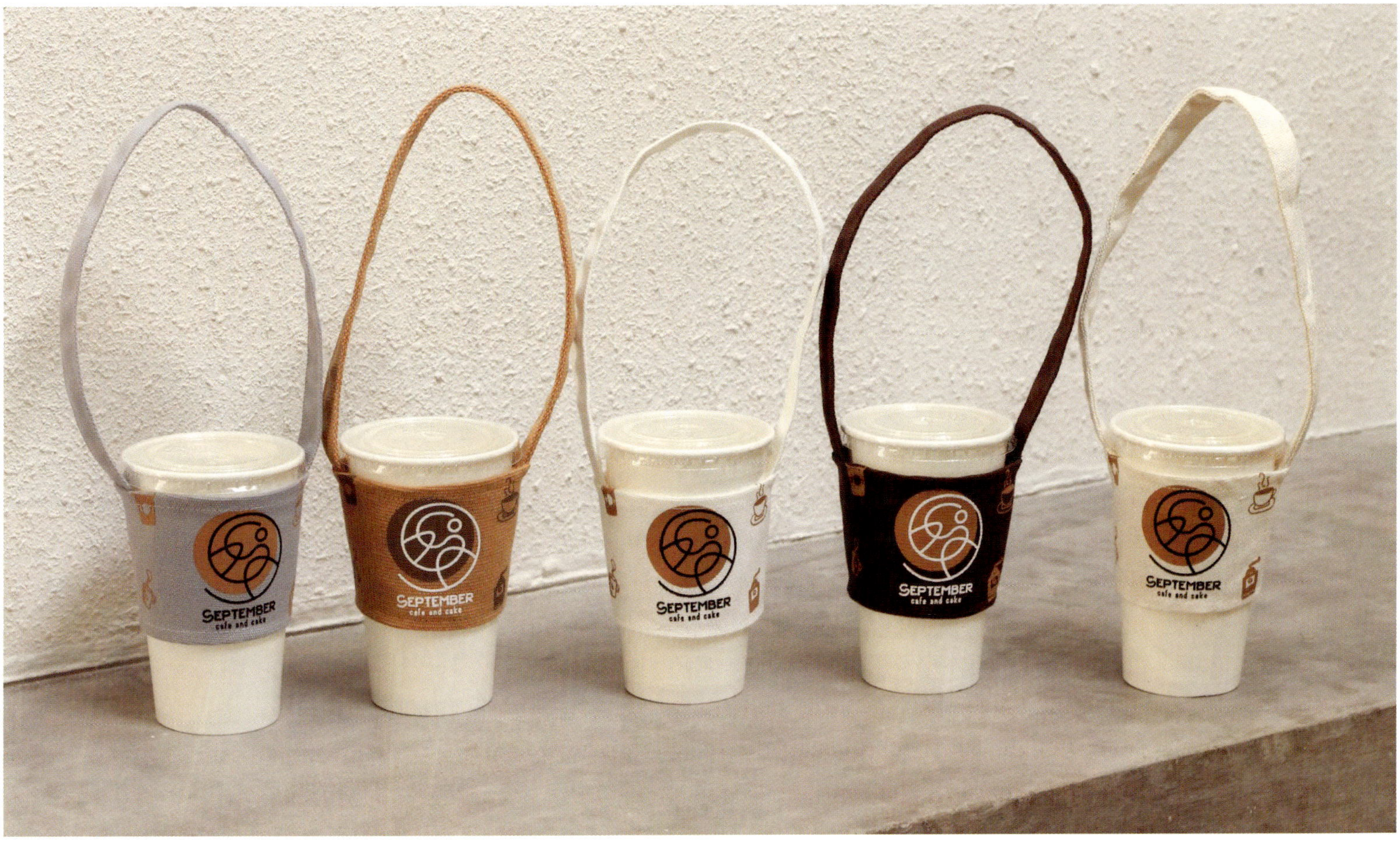
SEPTEMBER
cafe and cake

SEPTEMBER
cafe and cake

holder to handle
SEPTEMBER
cafe and cake
SEPTEMBER
cafe and cake

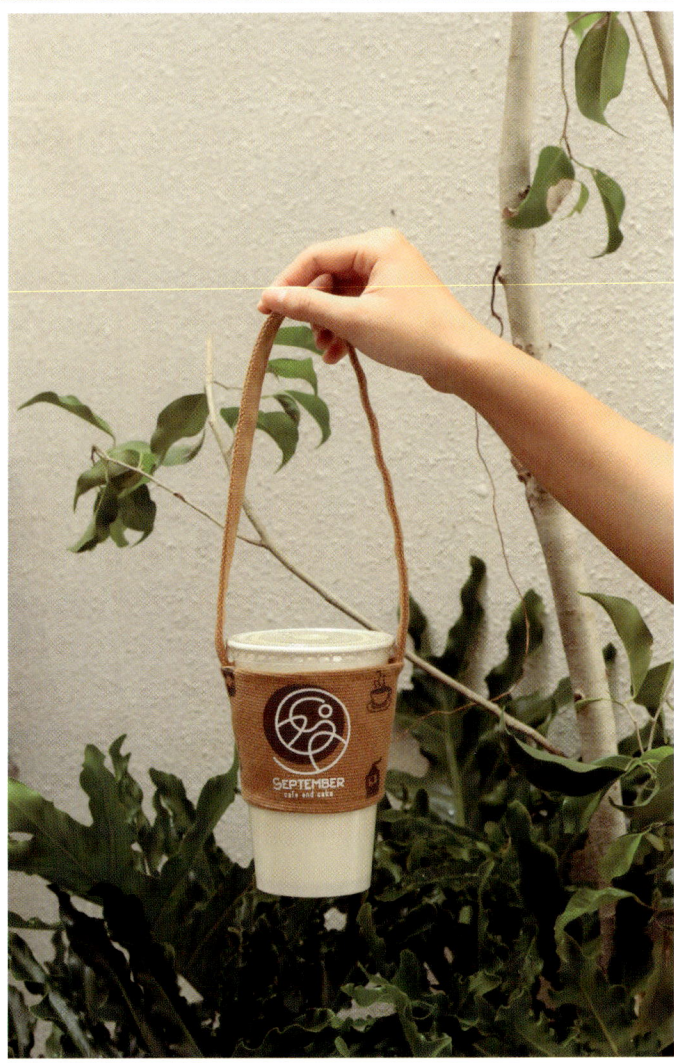

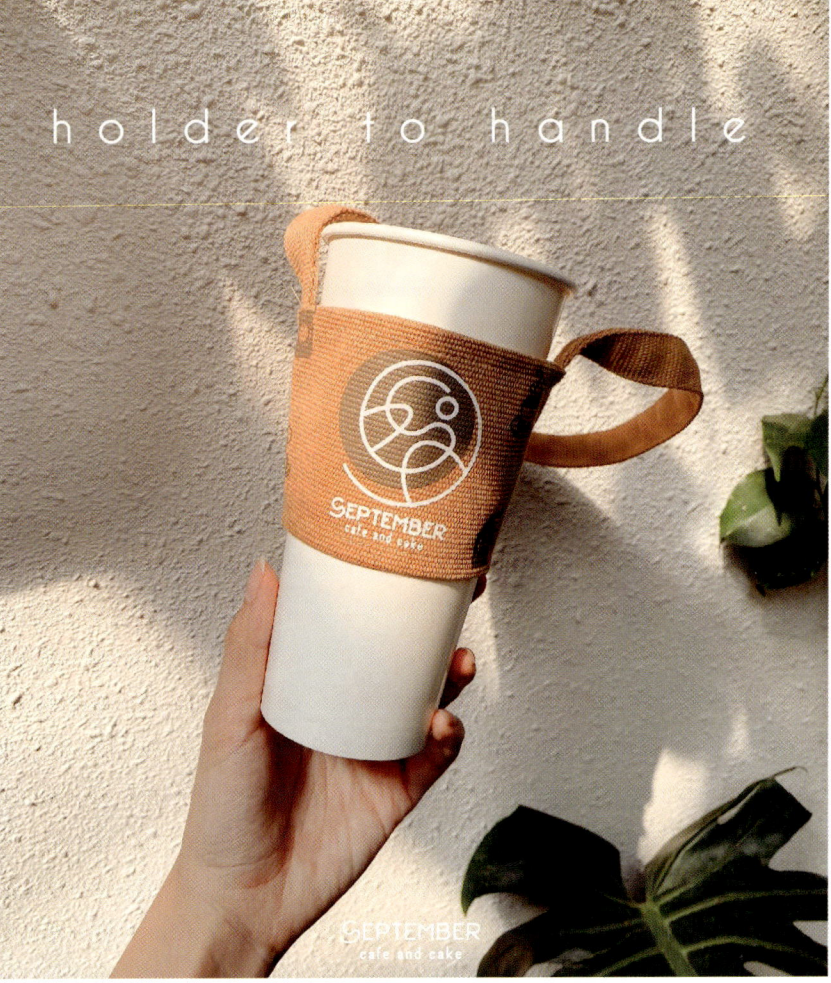

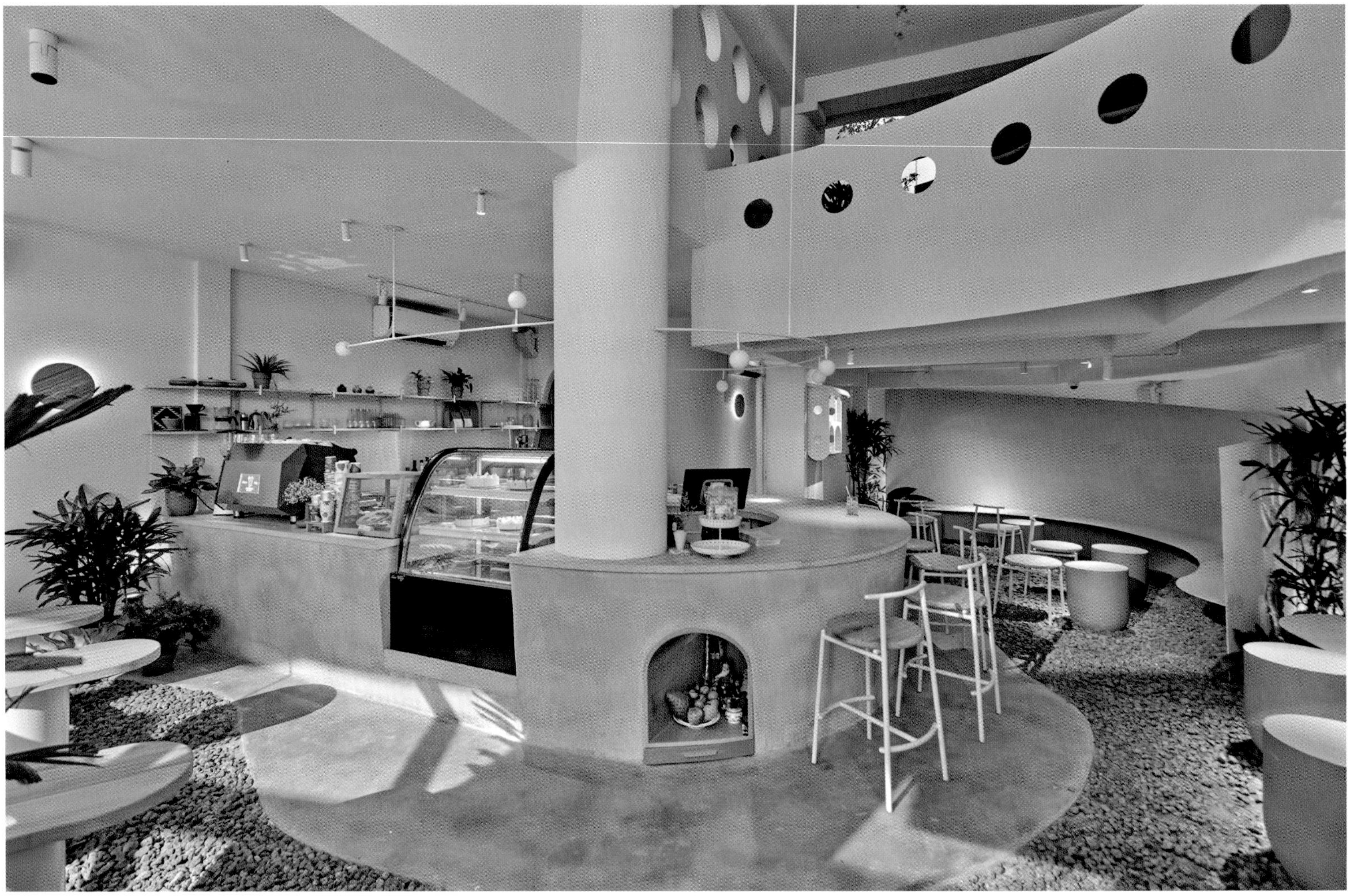

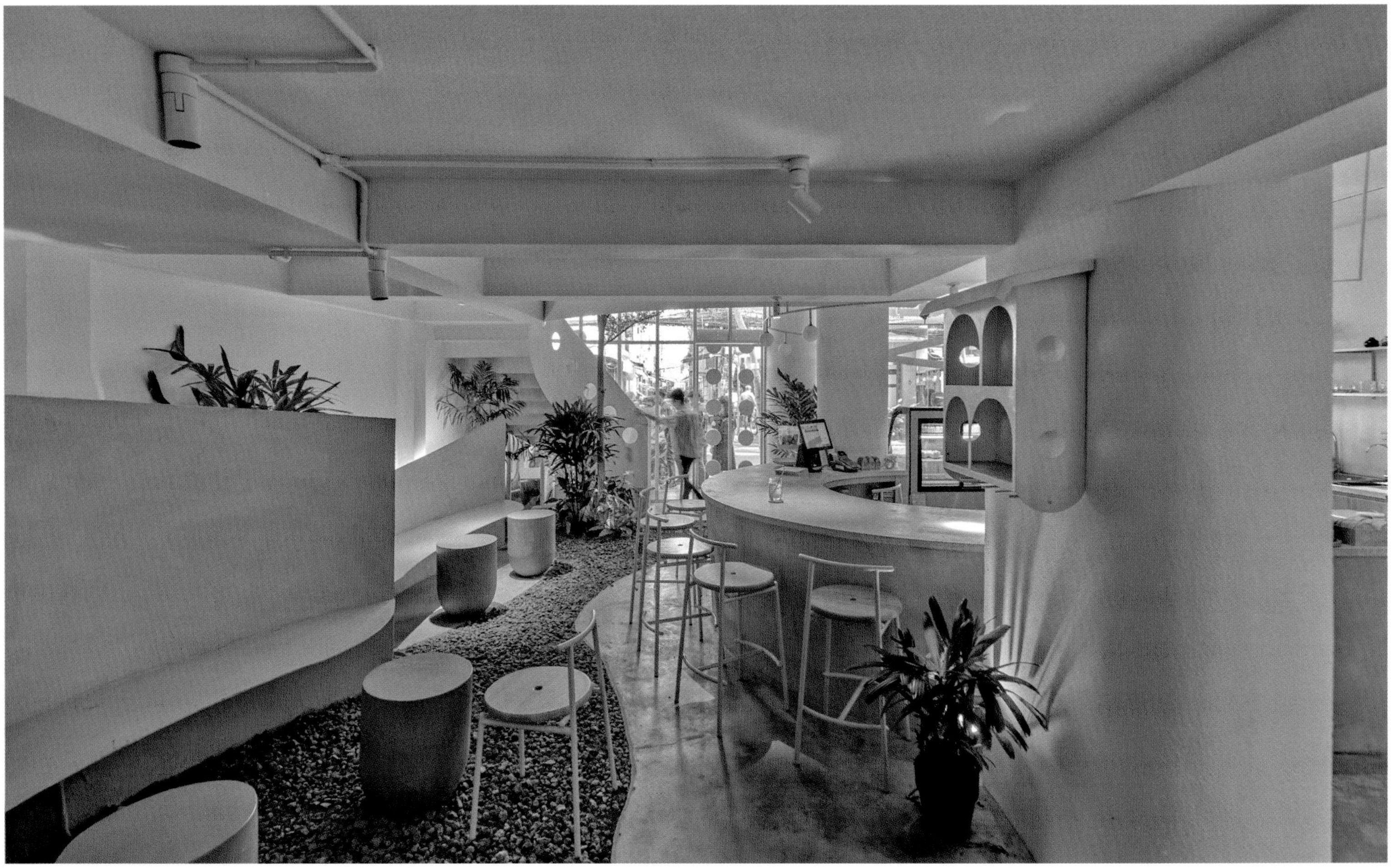

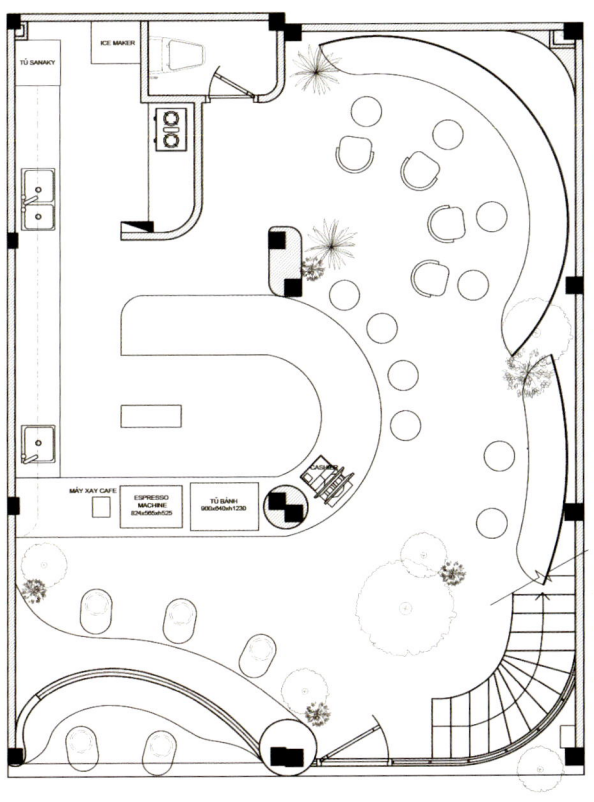
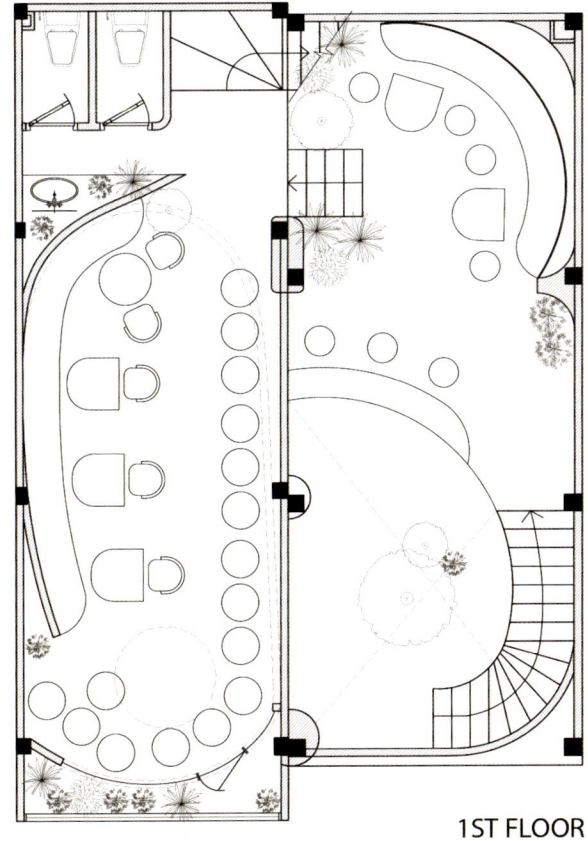

G FLOOR   1ST FLOOR

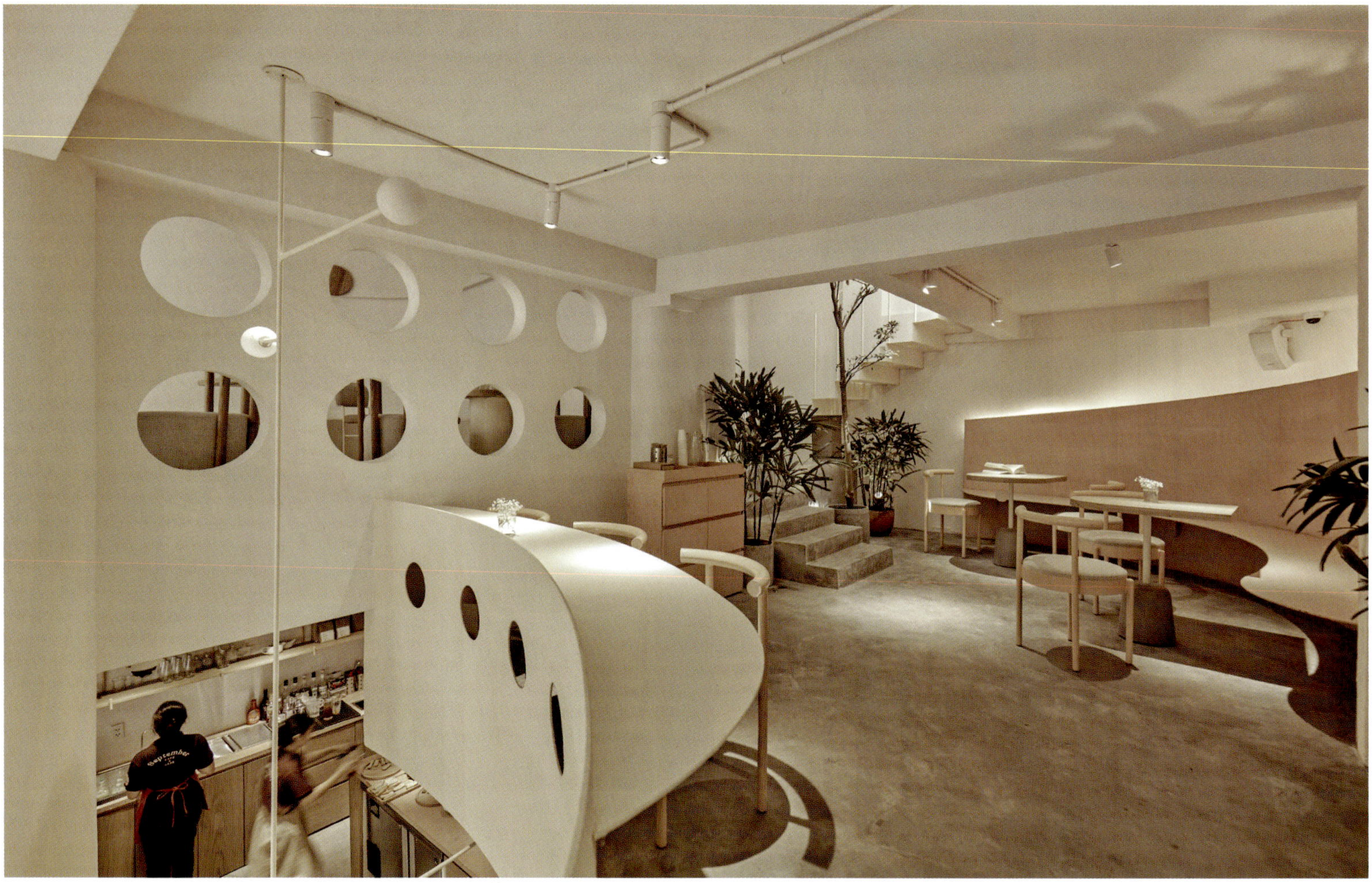

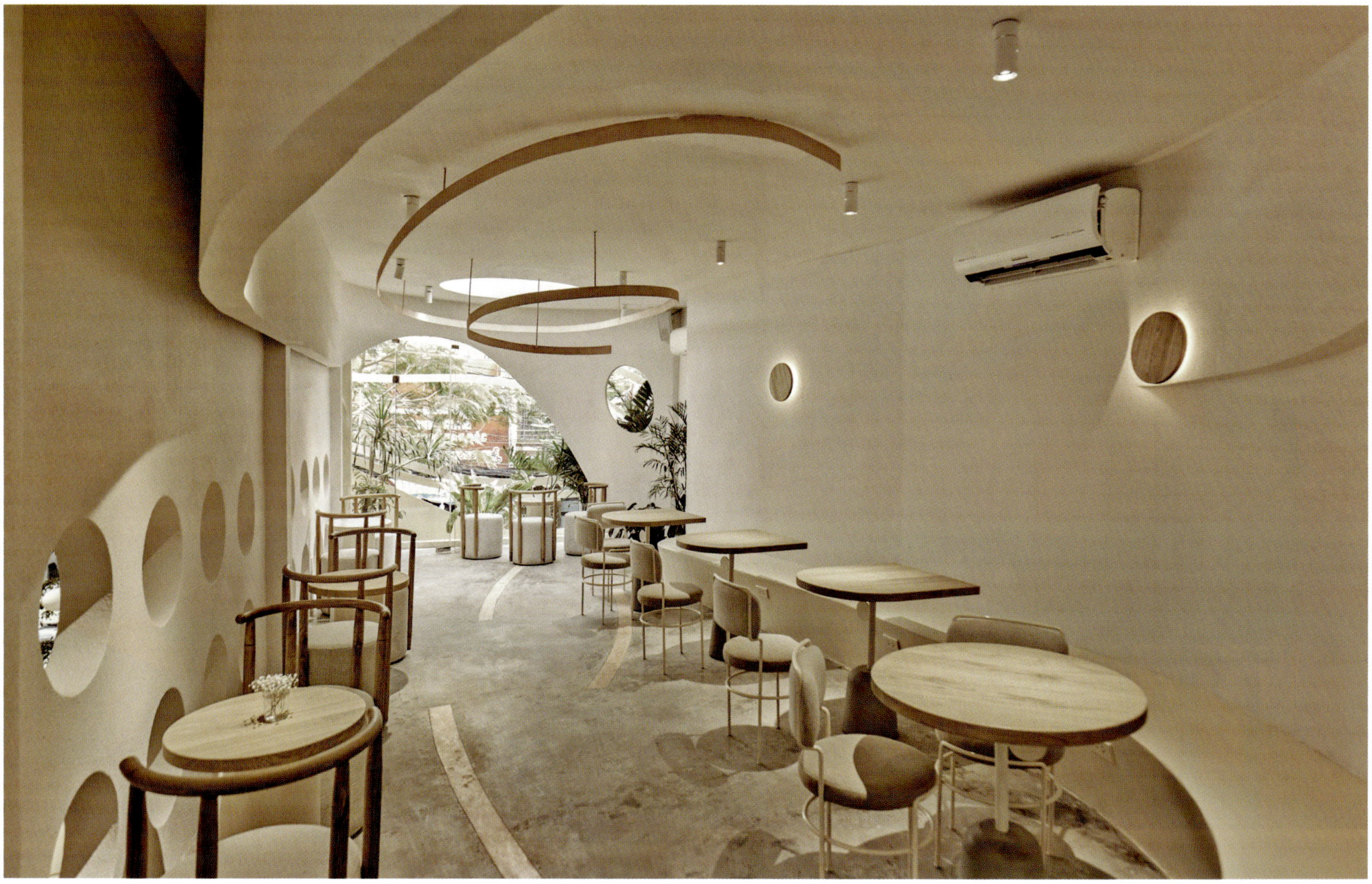

265

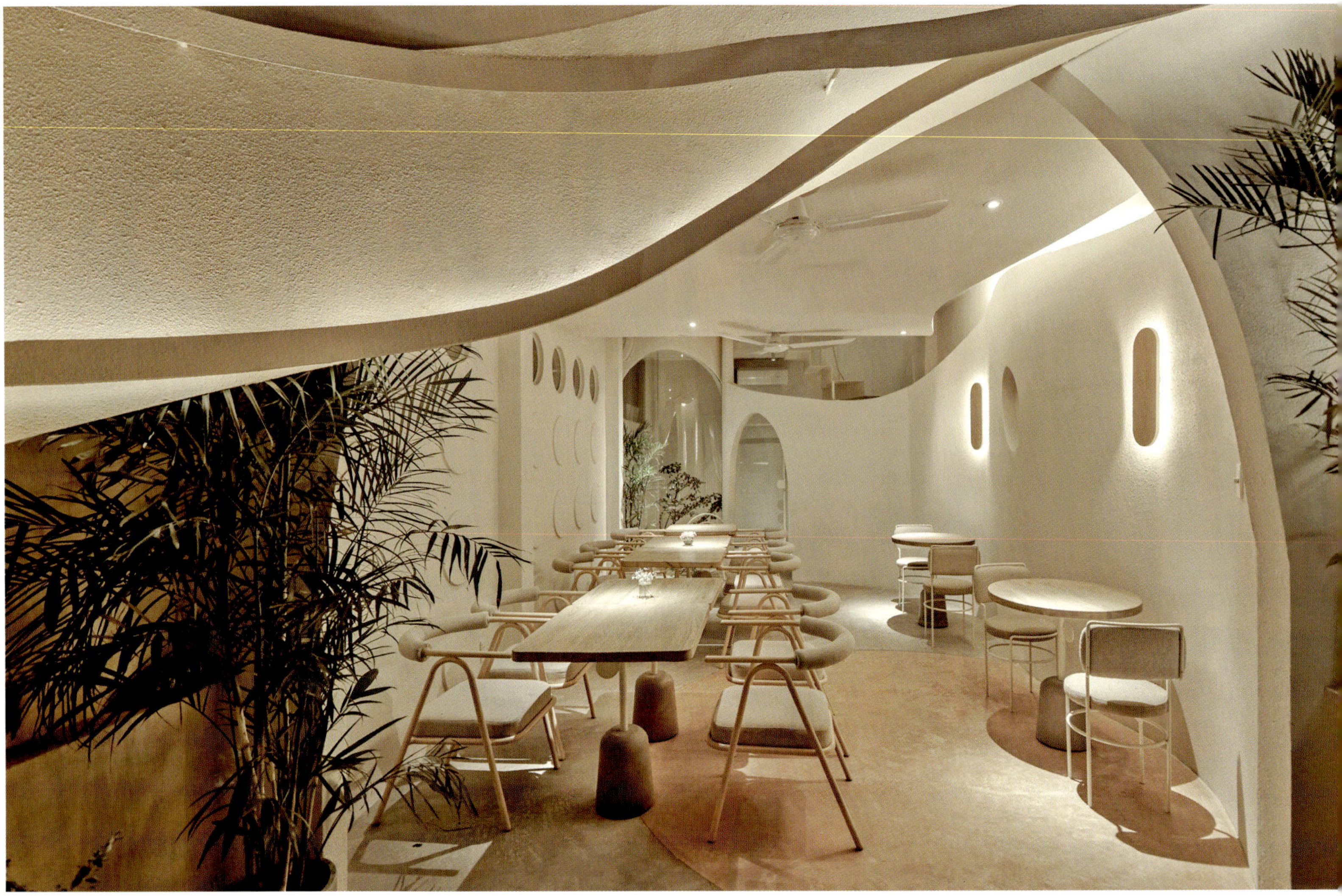

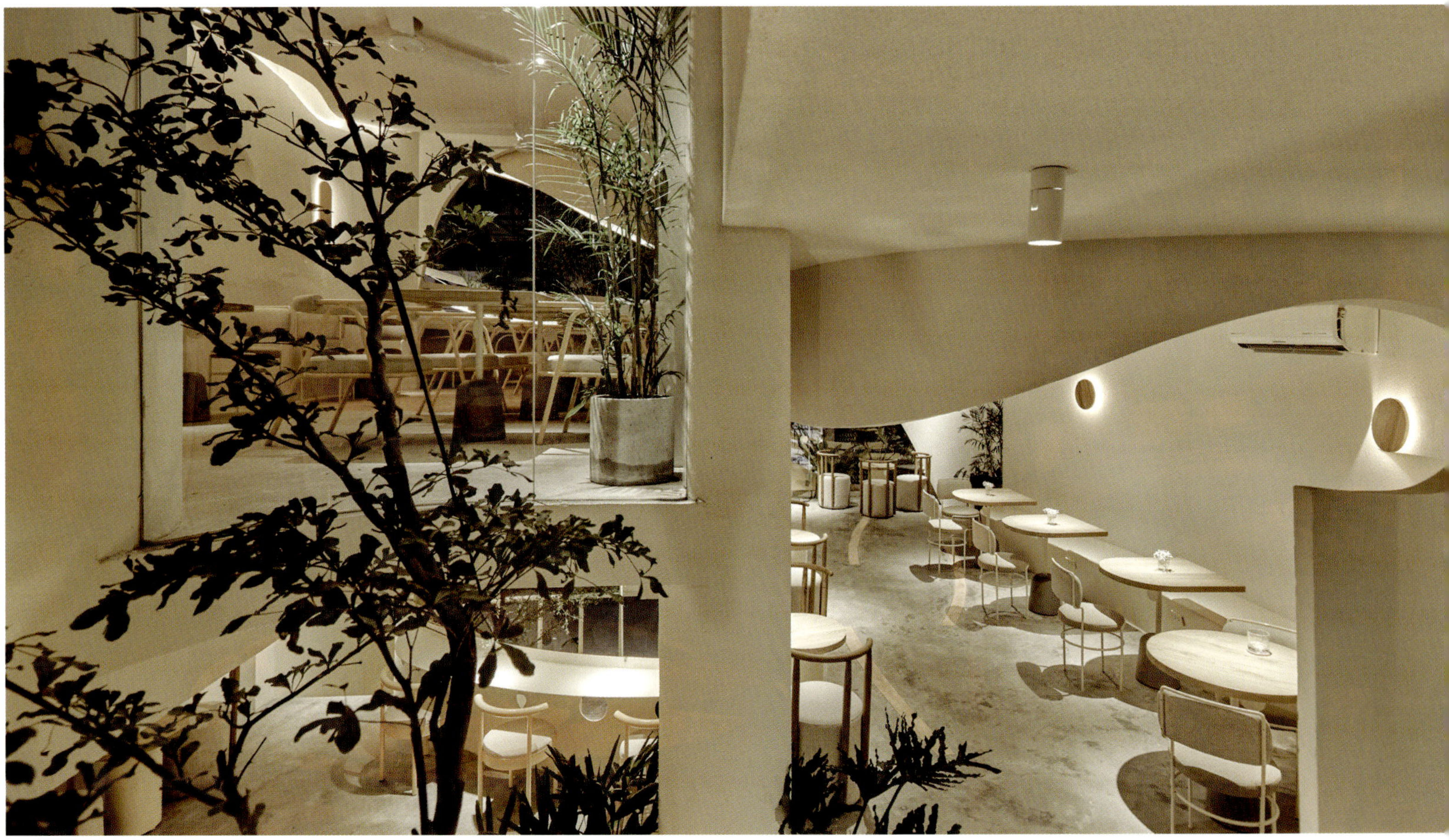

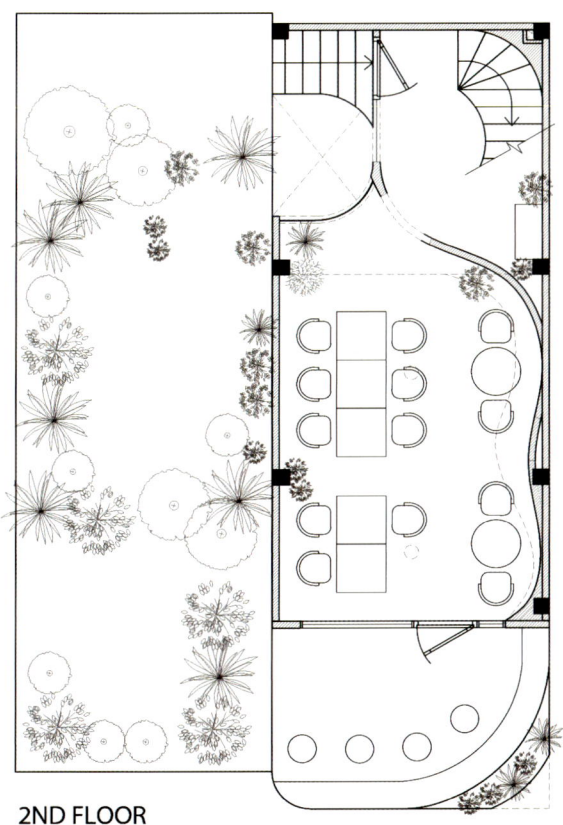

2ND FLOOR

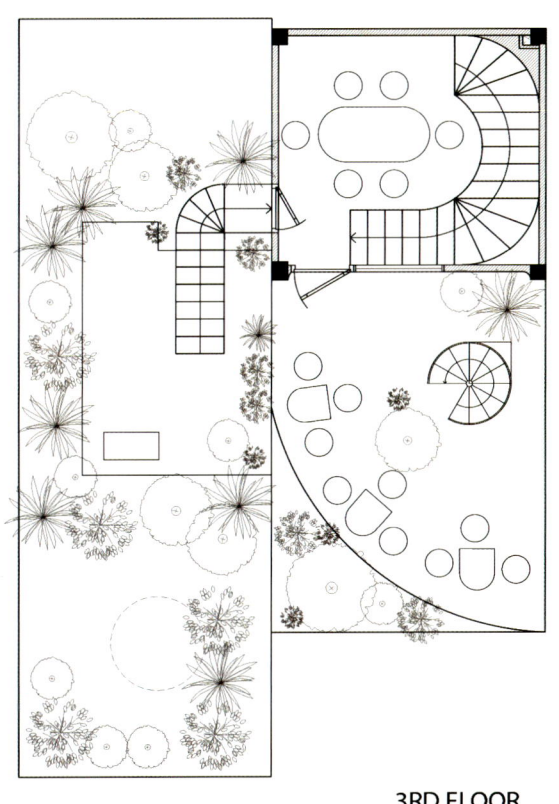

3RD FLOOR

Melrose Coffee
V46 Elgin Street, Central, Hong Kong

MELROSE Coffee

House of Forme?

House of Forme is a full service design and creative agency supporting the evolution of brands from initial concept to end user experience- Through a narrative based holistic approach. We design compelling stories, systematic identities, immersive environments and considered objects. We craft unique communications that addresses multiple layers of the human experience. Experiences that change the way people feel, think and do. Our designs are defiantly different. Tastefully executed and driven by strong narratives. We go beyond addressing relevance and commercial needs, beyond hollow pretty things, but compelling identities with a sip of nostalgia.

Design: House of Forme **Homepage:** https://www.houseofforme.com **Photography:** Vizion Creation
Location: 46 Elgin Street, Central, Hong Kong **Area:** 110 m²

California Dreamin' Melrose Coffee is a Los Angeles-themed cafe located in Soho, the destination for trendy eateries and bars. Serving nourishing green bowls and American-style pancakes, Melrose Coffee's Californian breakfast items recall the sunny romance and nostalgia of 1970's L.A. From the signature gradients on their collaterals to the half-mooned wall scones and curved facade, Melrose Coffee invokes the city's infamous sunsets at every turn. House of Forme worked with the client on their brand concept, visual identity, collateral development, interior design and FF&E design.

Sunset Cruising Down Melrose Avenue Carved from the classic shapes of mid-century signages and vintage postcards, the various logotypes for Melrose Coffee invoke a contemporary twist on Old Hollywood glamor, modernised by its application and vivid palette. Paired with primary typefaces made of classic serif fonts, the brand's identity is executed across its custom-cut and embossed collaterals, infused with warm gradients of romantic pinks, sunset yellows, and sandy beiges. Melrose Coffee delivers a glamorous, yet free-spirited L.A. lifestyle to the streets of Hong Kong.

Inspired by All-American Diners The brand identity of Melrose Coffee was extended across its interior design, with specific reference to Californian beach culture and classic American diners. Given the high yet narrow structure of the shop, we installed tilted mirrors for an illusion of depth and a false ceiling of light boxes that arch overhead, giving the space its sunny California glow. The bar was made with customised terrazzo and outdoor tiling meant for rocky exteriors, with striped details inspired by retro diner tables.

Our bespoke designs delivered both functionality and aesthetic in playful detail across the checkered quartz stone flooring, rounded dowel walls, and poolside benches that invite a stunning entrance into the coffee and brunch culture of the City of Angels.

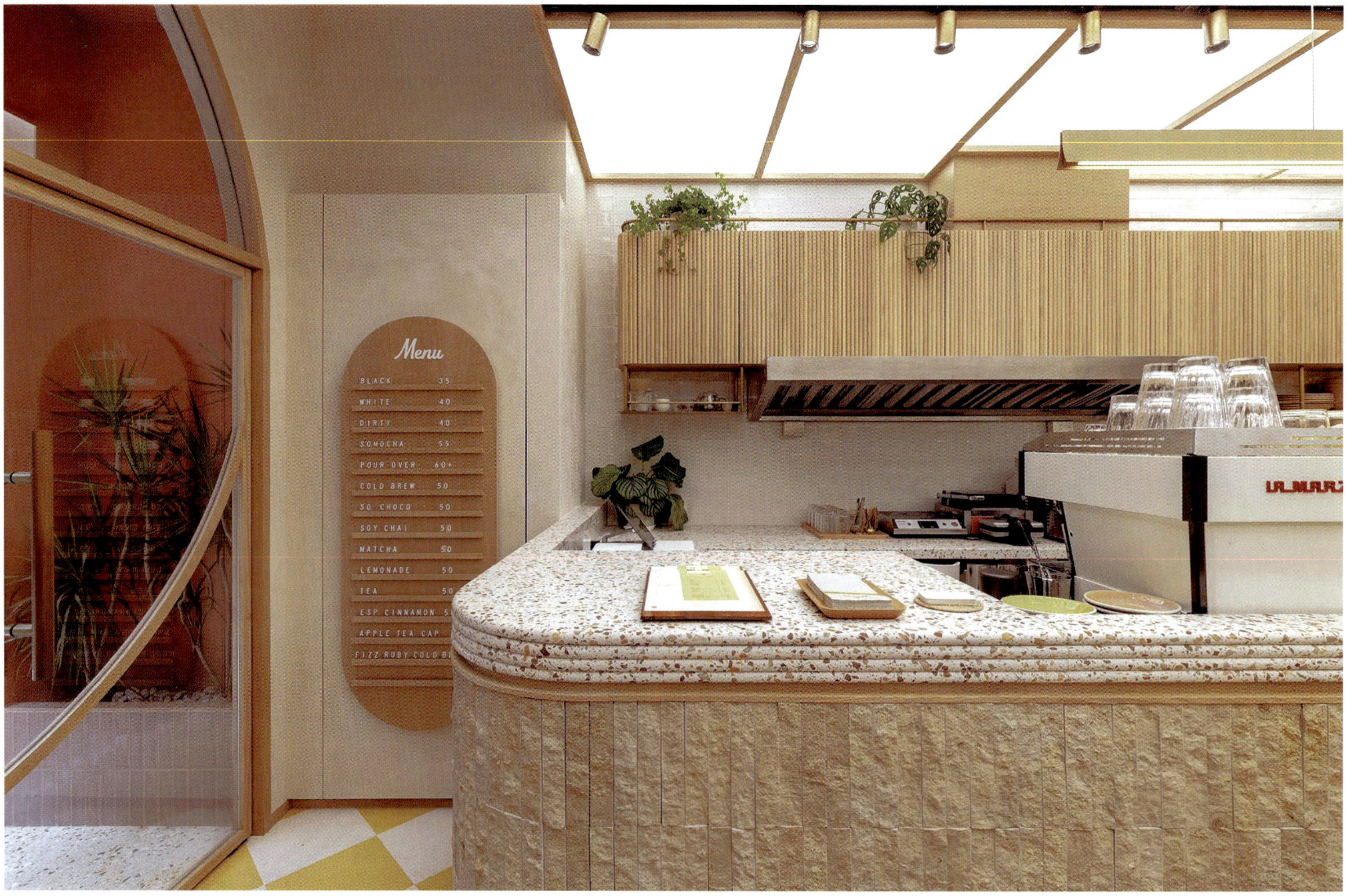
Menu
BLACK 35
WHITE 40
DIRTY 40
SOMOCHA 55
POUR OVER 60+
COLD BREW 50
SO CHOCO 50
SOY CHAI 50
MATCHA 50
LEMONADE 50
TEA 50
ESP CINNAMON
APPLE TEA CAP

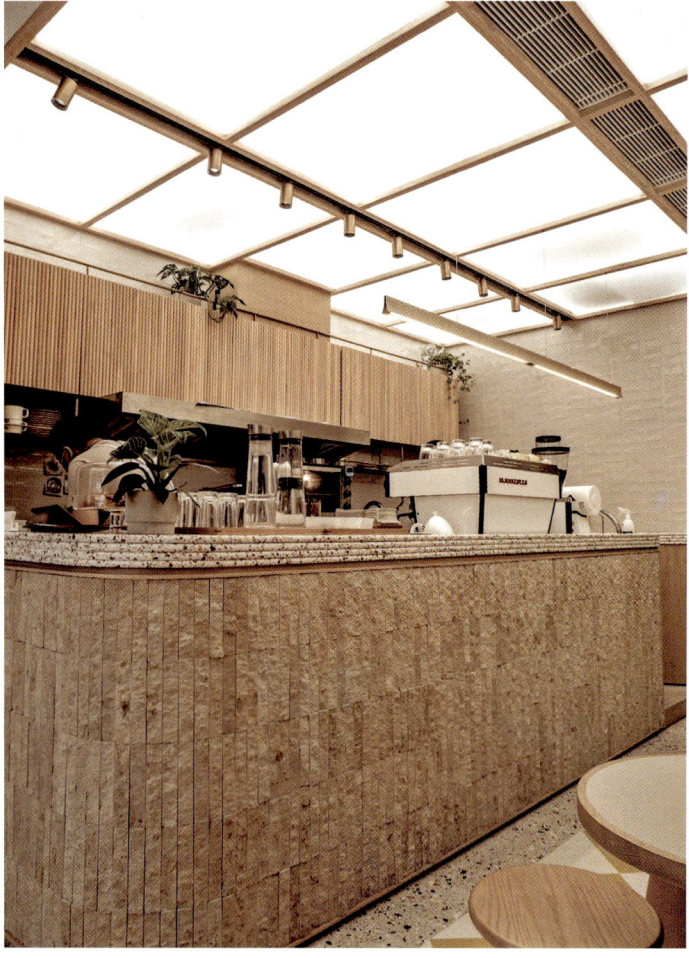

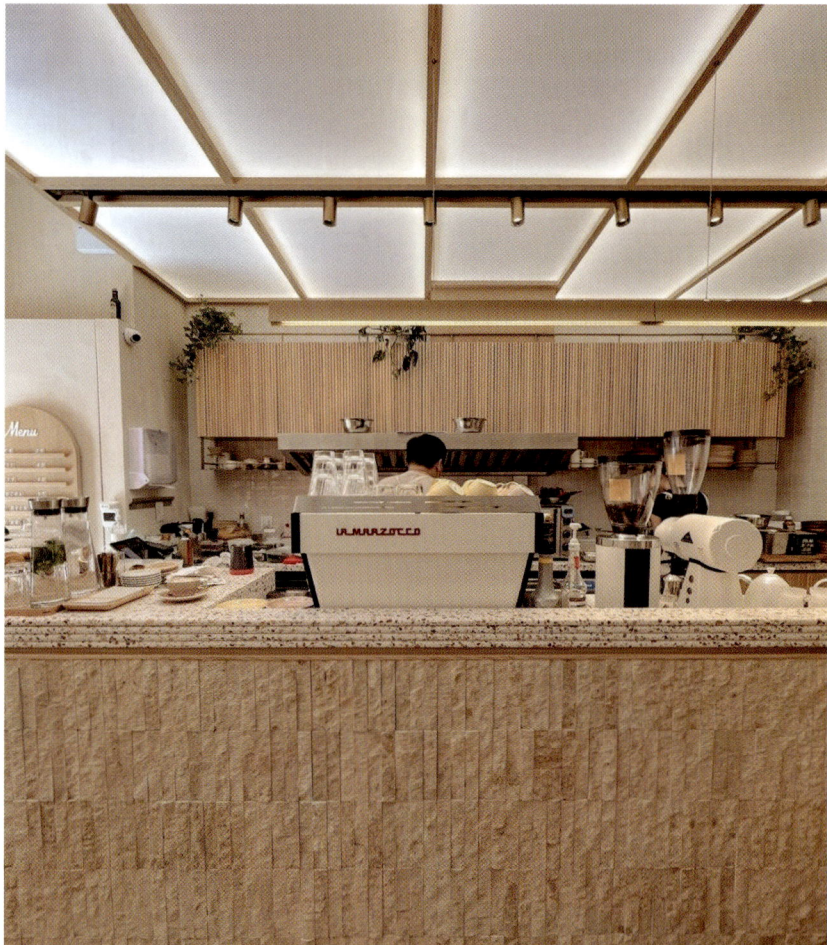

Abbocca Bistro Cafe
Beijing, China

abboCCa

Ramoprimo Architects ?

TUDIO RAMOPRIMO is a Beijing based architectural design practice founded by Italian architects Marcella Campa and Stefano Avesani in 2008 working in between Italy and China. The studio's name refers to the way they call in Venice some hidden lane and it's the Italian translation of the Chinese term Tou Tiao, which indicates the first lane of a series of Hutong alleys in old Beijing carrying the same progressive name. It marks the beginning of a newly born urban process. Our design projects range from urban planning to architecture, interior design and graphics. Main focus of the studio is the interaction between the social and the build environment offering at different scales vibrant design solutions for any contemporary urban living need.

Design: Ramoprimo Architects **Homepage:** https://ramoprimo.com **Photography:** Zhang Hui
Location: Beijing, China **Area:** 170 m² + 20 m²

Abbocca is an Italian bistro in the middle of the crowed Sanlitun area of Beijing. A large white curved diagonal ceiling, bright orange resin on floor and a touch of olive-green colour for both wine displays and food shelves define the space, specifically designed to convey a joyful atmospheare and feeling of being together. The narrow and low given space, has been open on a side to create a large facade to the street, where a concrete platform with solid seating settings is offering a large inteactive space for wellcoming and socialization.

Around a month ago Sanlitun Courtyard No.4 saw a new addition to its already impressive array of restaurants with Italian eatery abboCCa opening its doors. The new concept comes fro mthe same team behind Italian wine bar and restaurant Buona Bocca: it's owned by Fleur Xu and her Italian husband Alex Sanna, and managed by fellow Italian Alessia Dal Borgo.

AbboCCa serves up their food with a twist: there is no menu in the traditional sense. Instead customers select what they want from their two fresh food counters in an alimentari style. Now, given the current in-house dining ban it isn't possible to go to abboCCa but they're offering their same great food selection on delivery apps, and they truly have some unique options.

The most special thing about abboCCa, without a doubt, is that they make their own mozzarella in house. You can't get fresher than that! They are the only restaurant in Beijing to be doing this and it took them months of hard work and practice to create the perfect formula. Why not try their Caprese Salad (RMB 68) to see just how good it is!

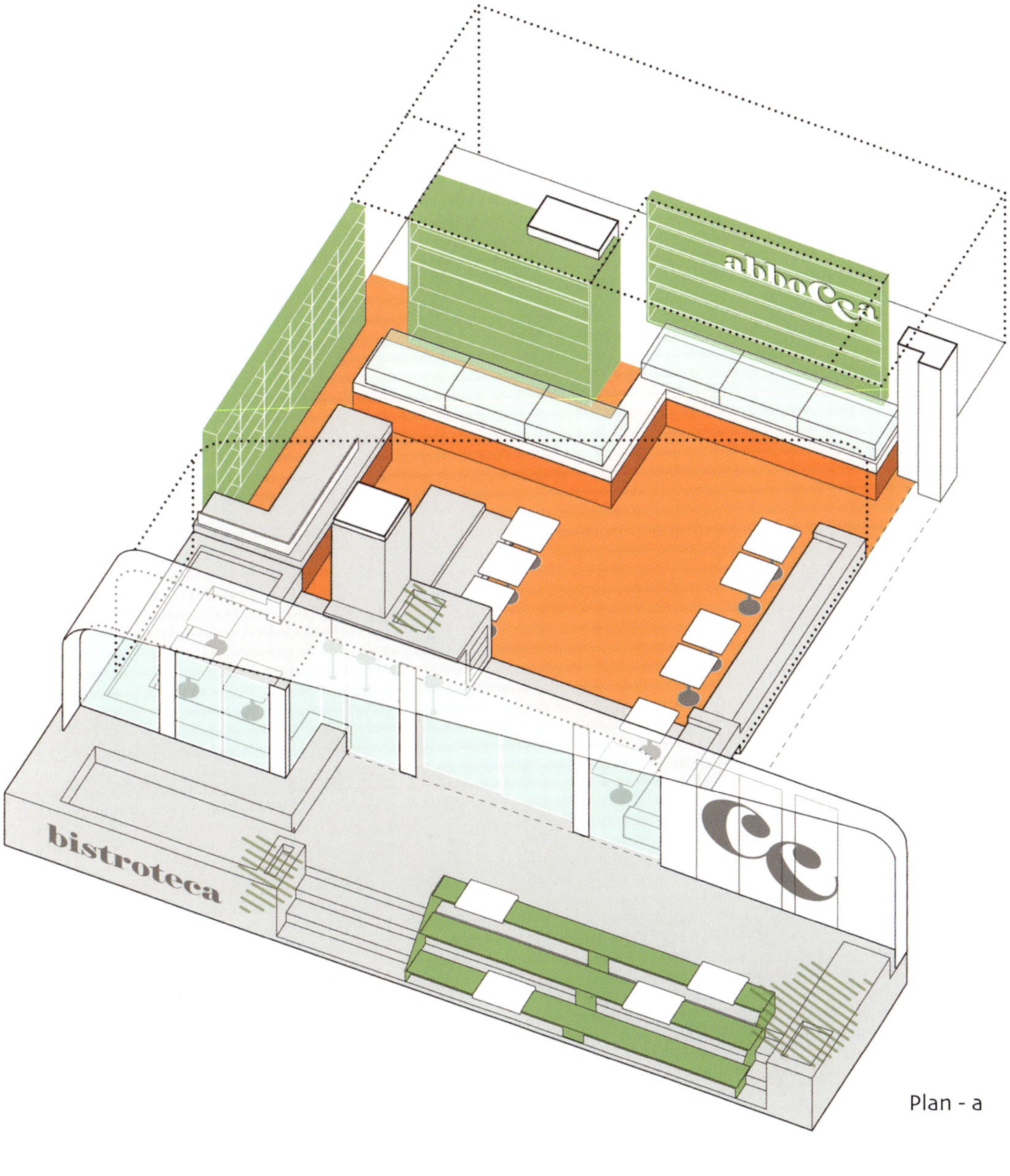

Plan - a

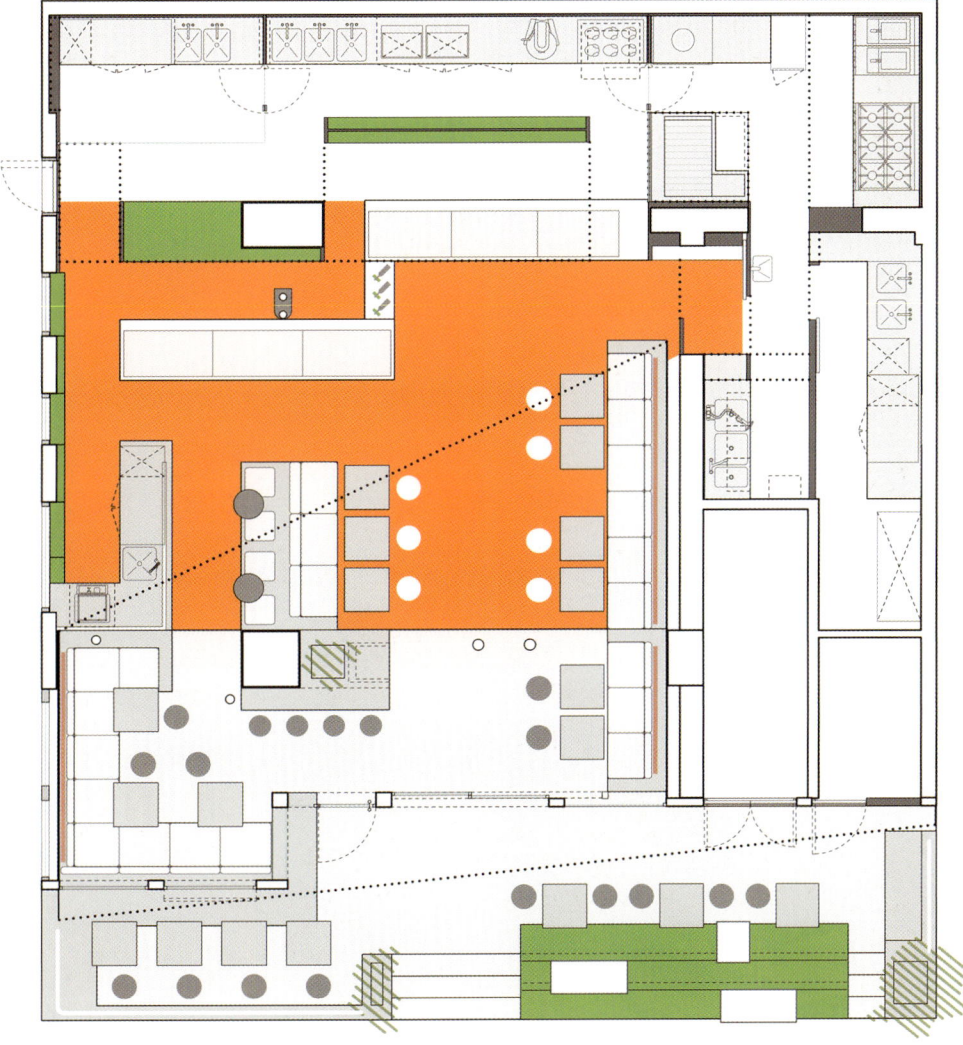

Plan - b

YAMA Coffee Shop
Thanh pho tay ninh, Vietnam

KSOUL Studio ?

KSOUL is a Saigon-based studio of architectural professionals driven to deliver progressive projects for ambitious clients.

Nguyen The Huynh(founder); Founded KSOUL studio and leads the design direction of the practice with a combination of his inquisitive nature, creativity and proven technical skill.

Design: Ksoul studio   **Homepage:** https://ksoul.studio/contact   **Photography:** Valor Studio
Location: Thanh pho tay ninh, Vietnam   **Area:** 195 m²

Greeting the two owners of Yama coffee brand – Ky and Phuong – Ksoul was surprised and energized with the remarkable power of youth and passion that they used to grow their products. Knowing that YAMA has been continuously putting effort to improve their beverage product quality and adapt to the taste of the locals really had us appreciate and respect them even more.

he design of this new location actually is inspired by the original brand recognition of YAMA – You Are My Angel, picking up some interesting additional figures while attempting to keep the images that has been familiar to the majority of their supporters, producing a fresh and tight connection between the phases of their development. Following the fact that Tay Ninh is in its modernization, Ksoul decides to use the cement gray coloration which represents the urbanization concept, and yellow which is the brand color and also the reminder for the sun that has shined on this land. It is also safe to say that this location could be the beginning of the streak of luck for YAMA because yellow happens to be the most trending color in 2021.

In order to create a refreshing industrial-oriented space and impressive light effects, Ksoul has put in use the three significant materials which are cement, concrete, and stainless steel. The bar area is the first to be completed using cement and cladded with stainless steel that allow it to shine brightly to any customers as soon as they set foot into the store.

One special thing about this project is that a majority of interior furniture are customized and produced particularly for this brand. The stainless steel table out on the yard, the couch, the stool, or even the large desk on the second floor are meticulously measured and arranged to satisfy the structural and functional requirements. In addition, you would easily realize that the store is very spacious and comfortable thank to the lighting system application and the installation of glass tiles to help the place absorb natural light from the outside and highlight the yellow colored central area.

The yellow epoxy paint matches with the wall blocks color and the spiral staircase which lead to the second floor, delivering a breakthrough highlight but still linking to the YAMA images located around the space. The angel images, however, are not directly displayed to the brand lovers, but the image of the sun is put on some of the decoration spot to subliminally express the spiritual value of the angels and pay tribute to the supporters of YAMA who has been walking with them throughout their journey.

Ksoul is confident that we have made use of our creativity and understanding of the brand to deliver the most appropriate and impressive design to them. Not only do we focus on aesthetic value, but we also care about customers' experience and the effectiveness and profit that our project brings to them.

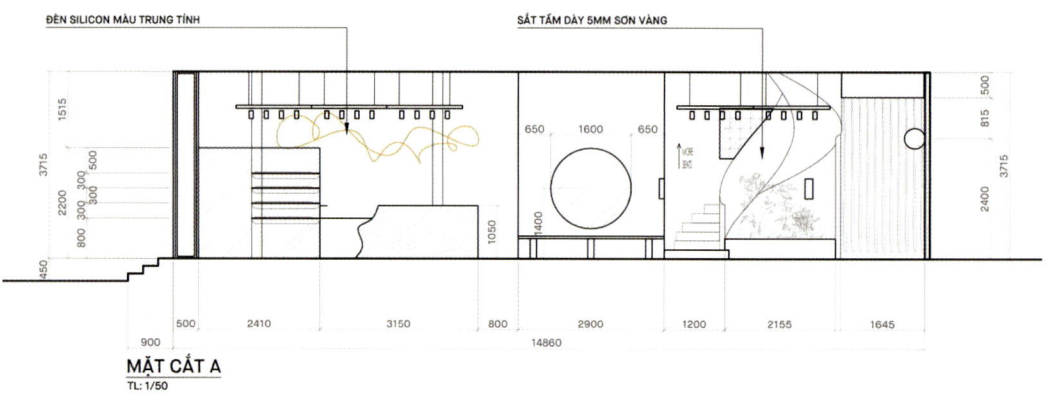

MẶT CẮT A
TL: 1/50

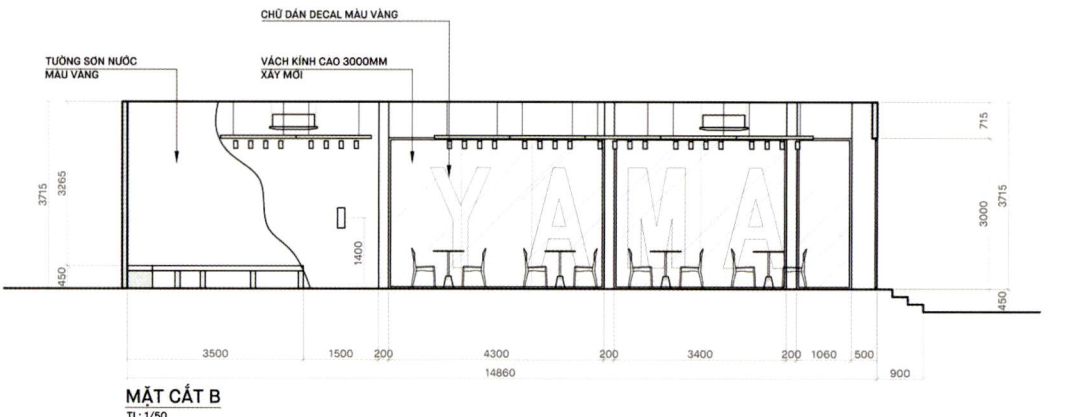

MẶT CẮT B
TL: 1/50

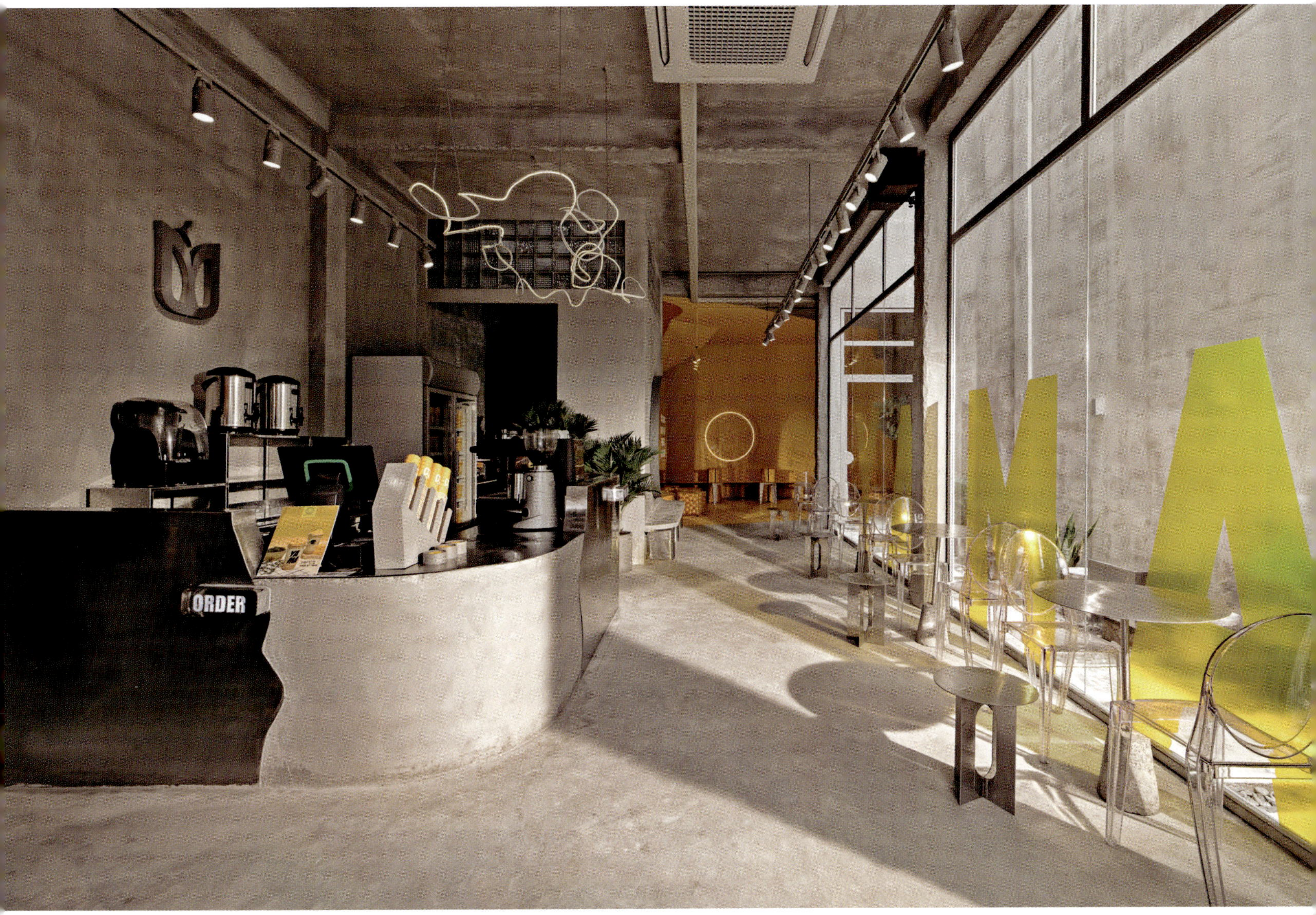
ORDER

YOU DON'T
HAVE TO BE
PERFECT
TO BE HERE

YOU ARE
MY ANGEL

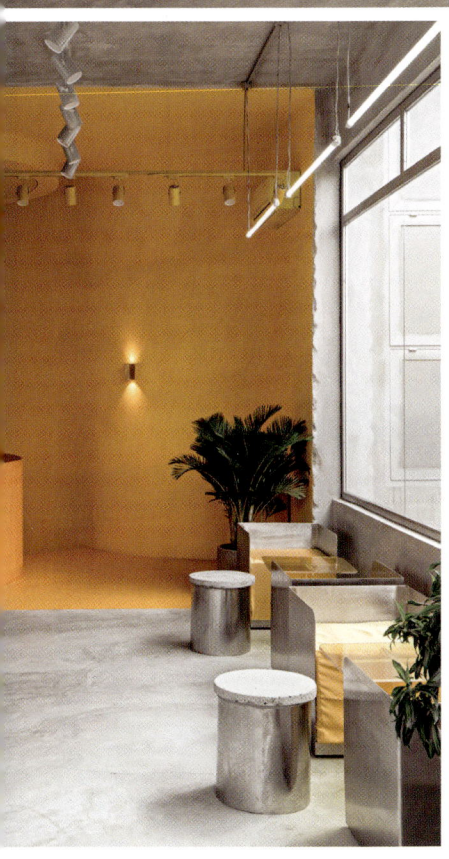

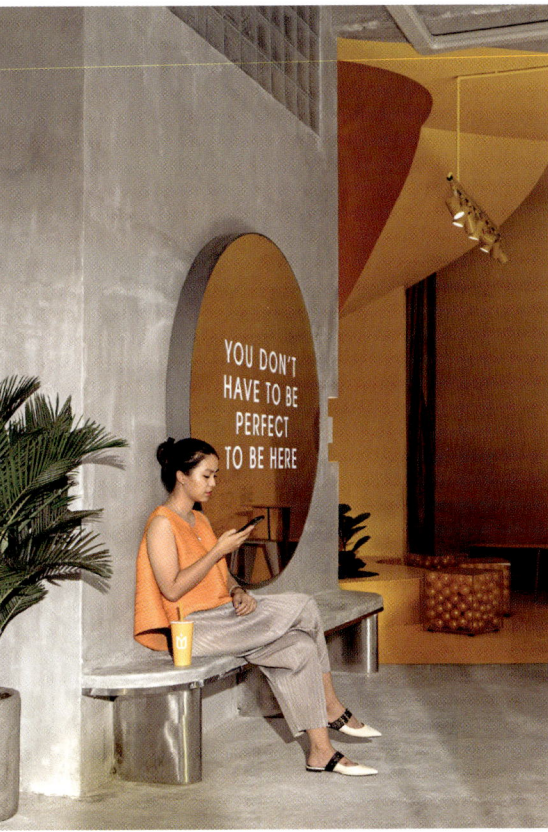

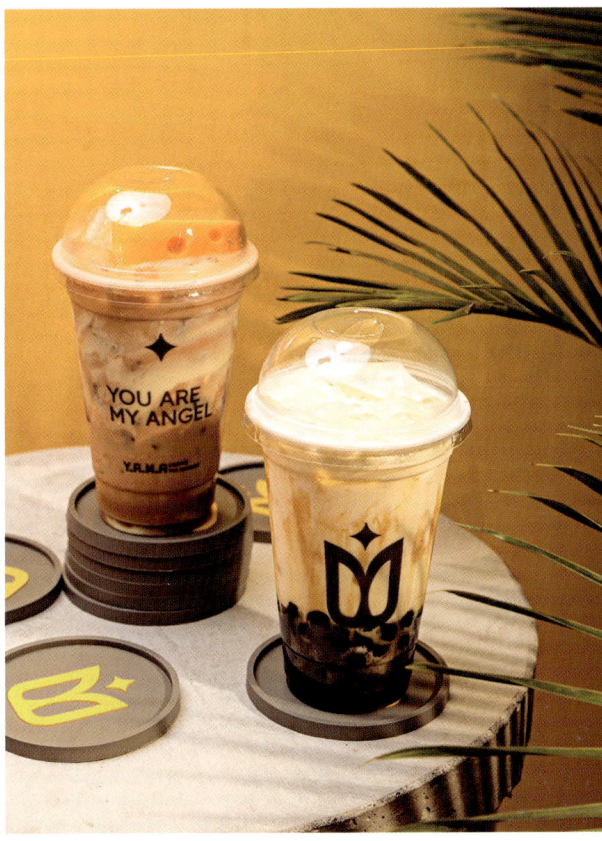

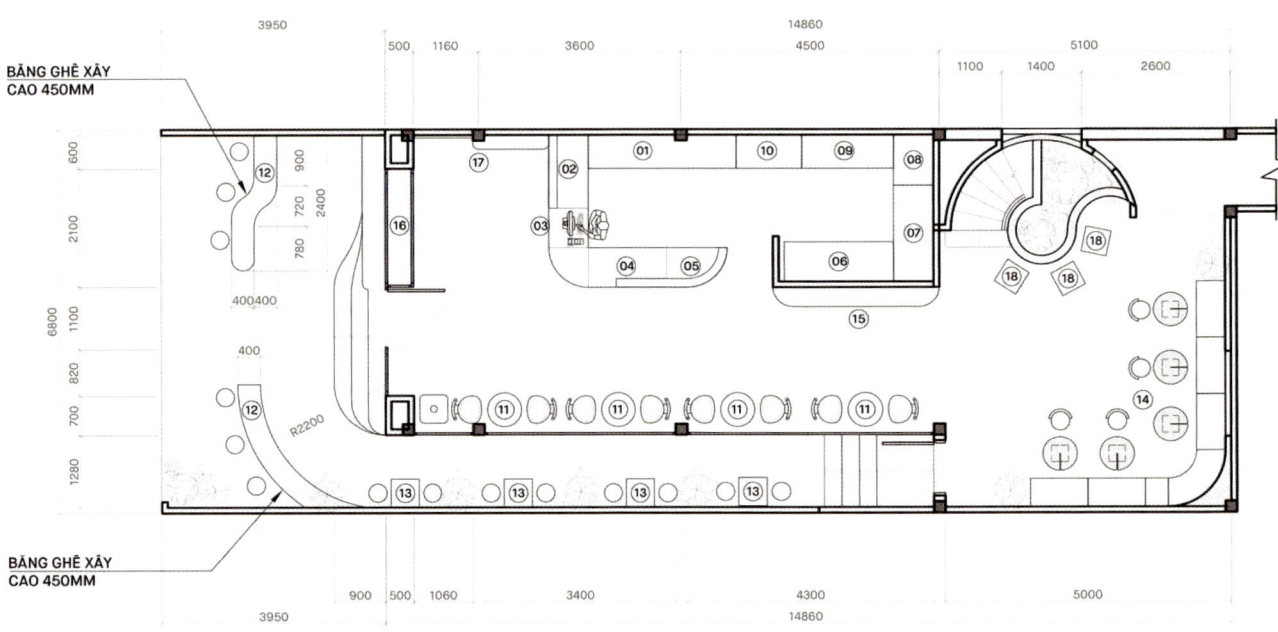

MẶT BẰNG BỐ TRÍ TẦNG TRỆT
TL: 1/50

01 KV PHA TRÀ SỮA
02 KV PHA ĐÁ XAY
03 KV ORDER
04 KV PHA CAFE + WAFFLE
05 PICK UP
06 BỒN RỬA + MÁY BINGSU
07 TỦ ĐÔNG
08 THÙNG ĐÁ
09 MÁY CREP + BẾP ĐIỆN
10 TỦ MÁT ĐÔI + BỒN RỬA
11 BÀN 2 NGƯỜI
12 BĂNG GHẾ NGOÀI TRỜI XÂY GẠCH CAO 450MM
13 BÀN 2 NGƯỜI NGOÀI TRỜI
14 BĂNG GHẾ TRONG NHÀ
15 GHẾ CHỜ
16 WINDOW DISPLAY
17 KỆ TRƯNG BÀY SẢN PHẨM
18 BÀN HỘP KÍNH

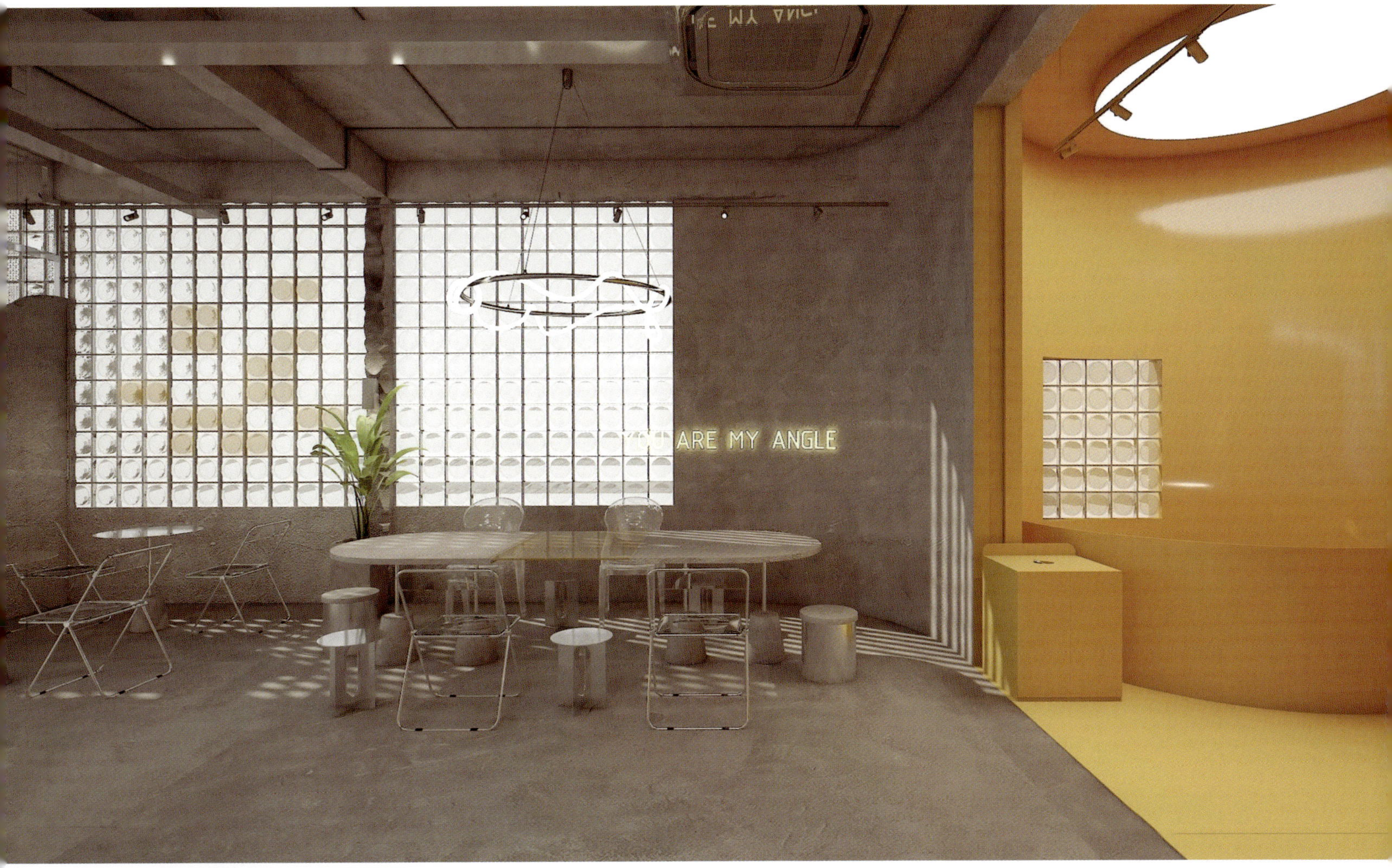

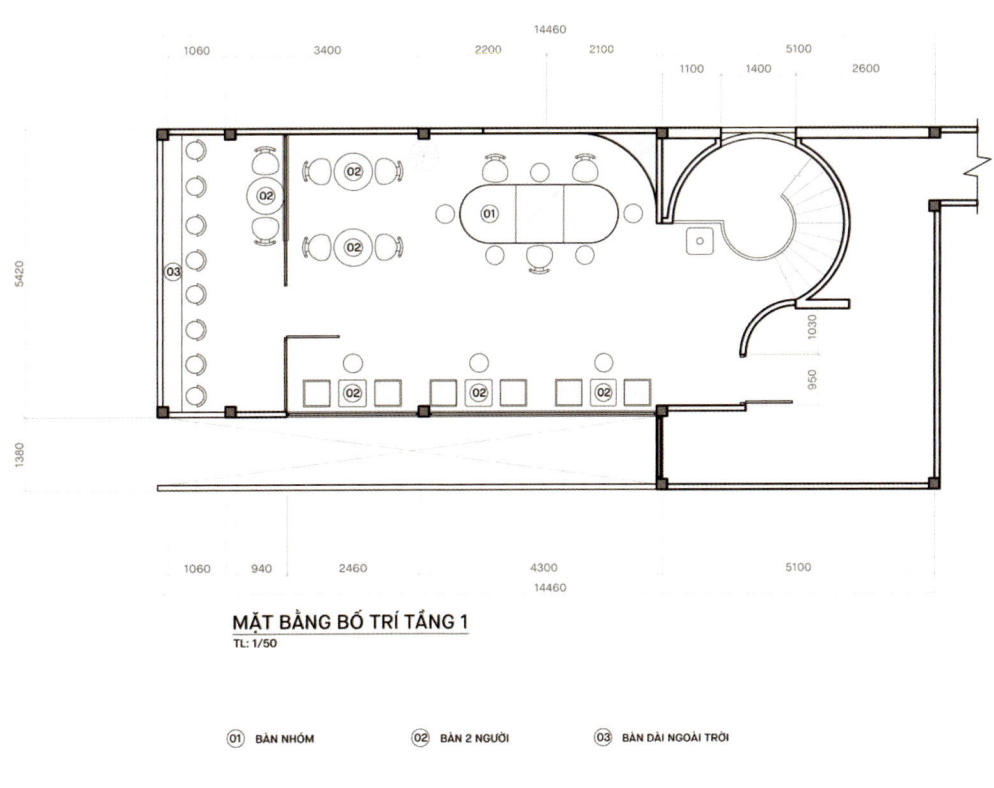

MẶT BẰNG BỐ TRÍ TẦNG 1
TL: 1/50

① BÀN NHÓM   ② BÀN 2 NGƯỜI   ③ BÀN DÀI NGOÀI TRỜI

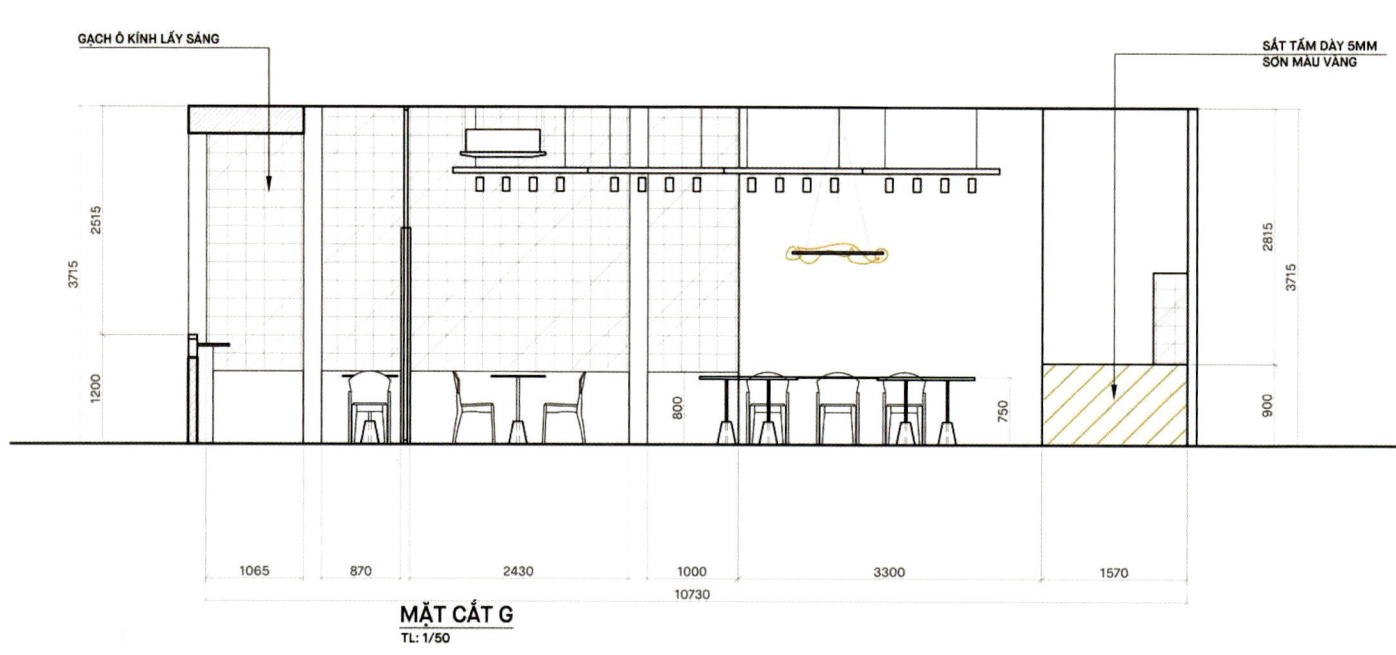

MẶT CẮT G
TL: 1/50

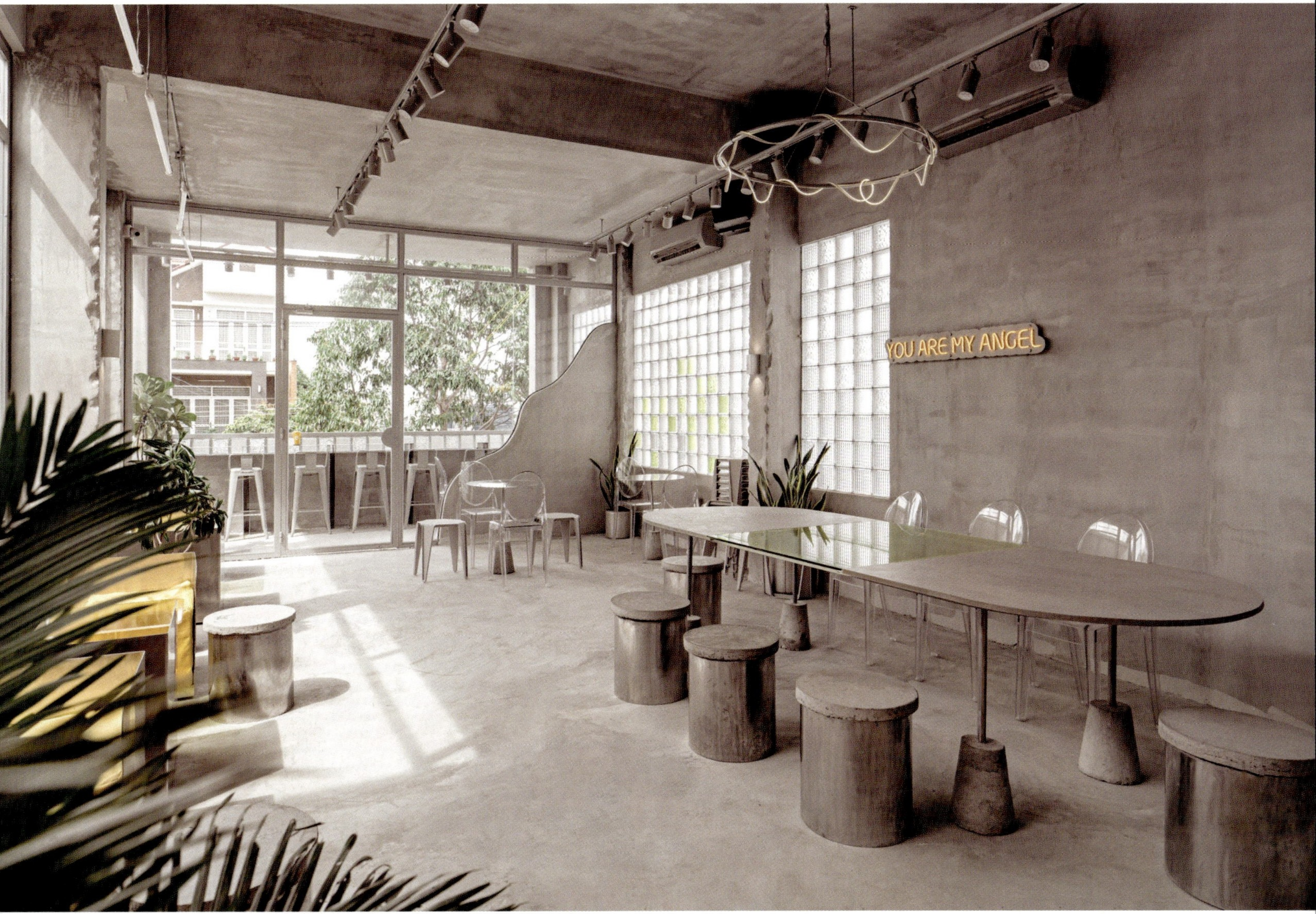
YOU ARE MY ANGEL

YOU ARE
MY ANGEL
CASADIO

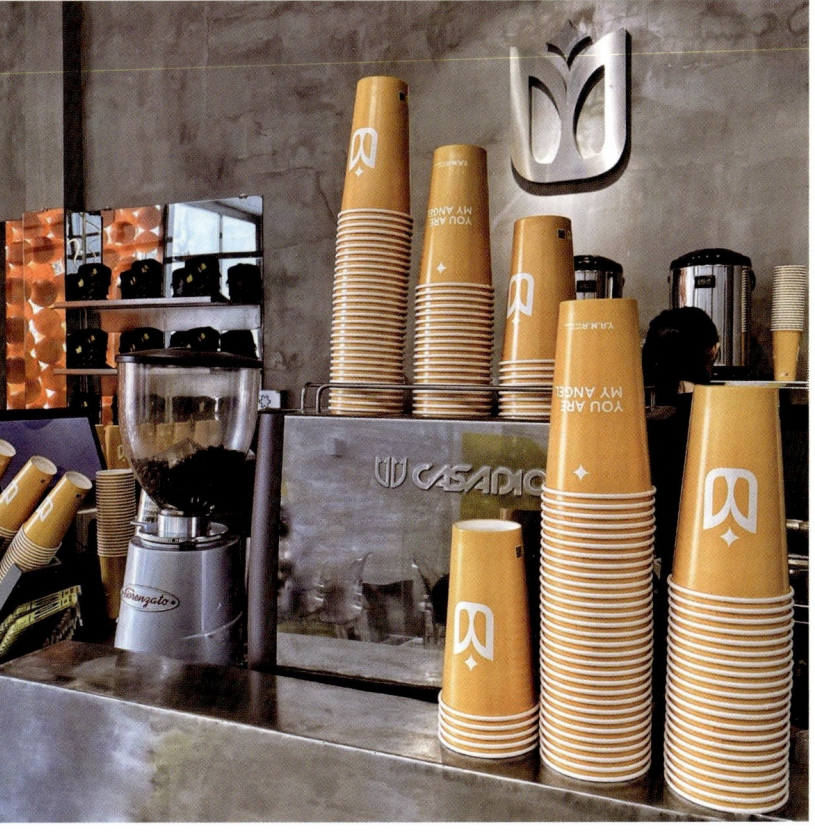

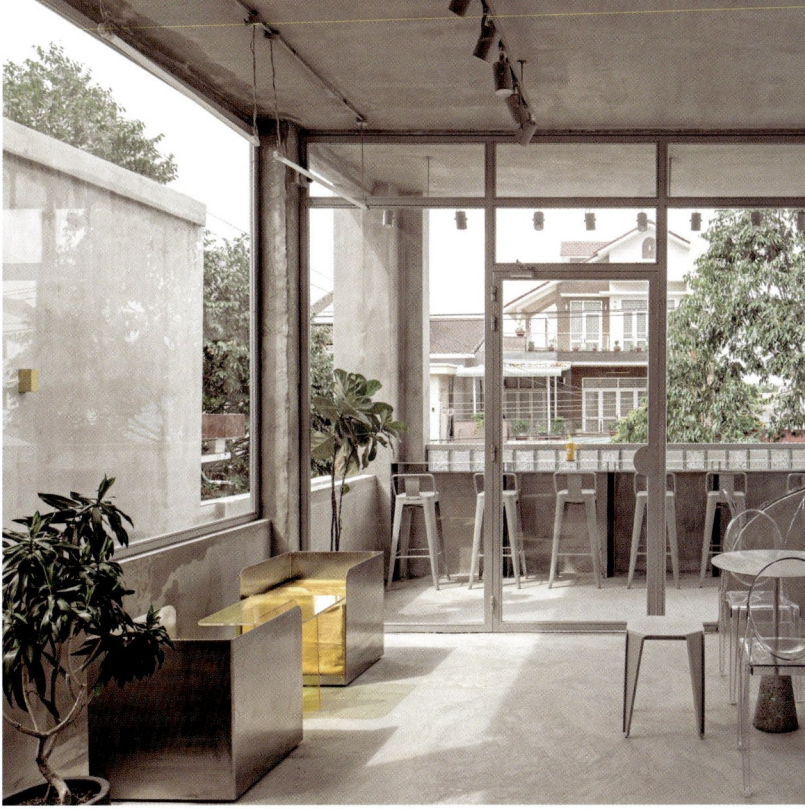

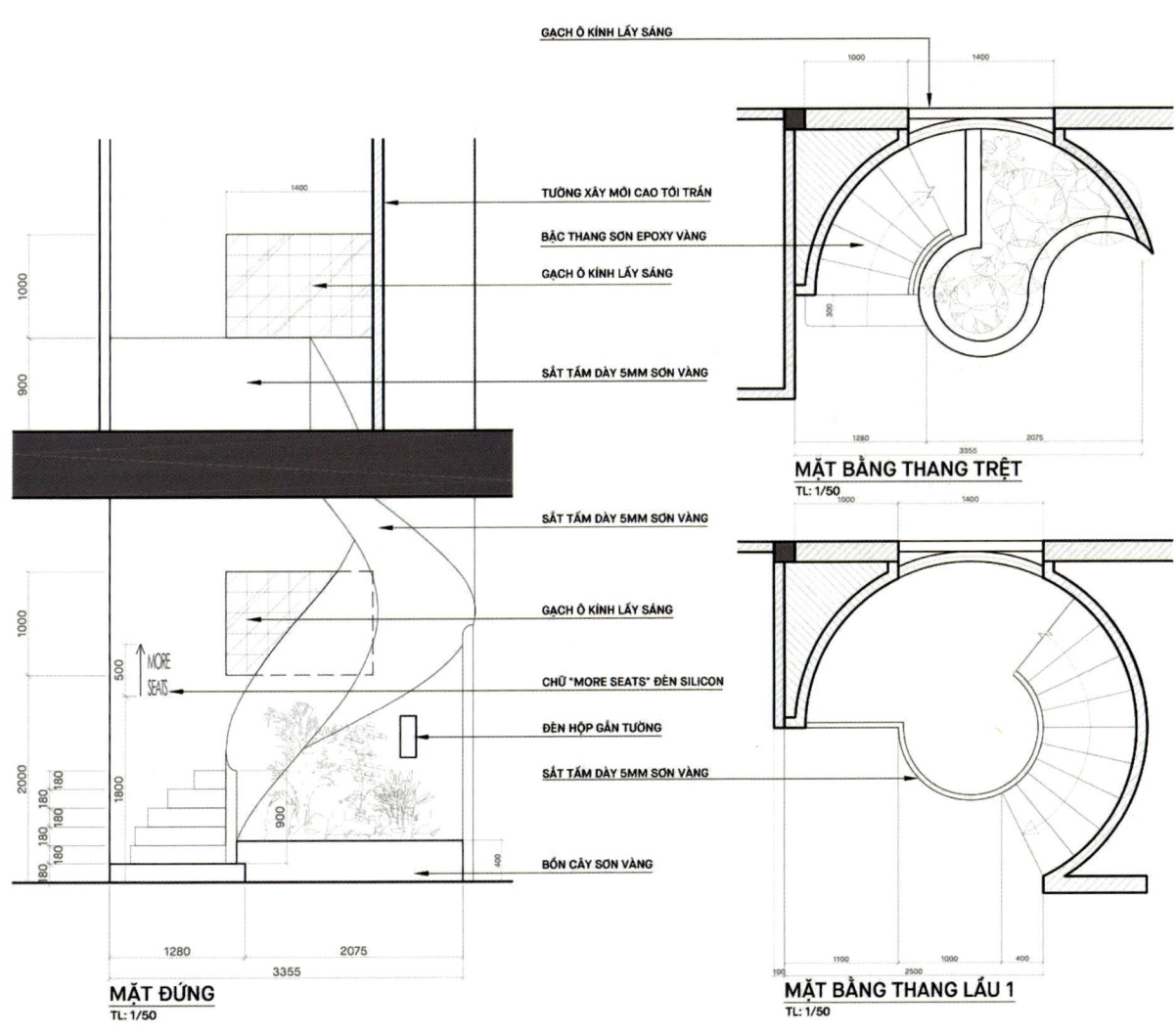

MẶT ĐỨNG
TL: 1/50

MẶT BẰNG THANG TRỆT
TL: 1/50

MẶT BẰNG THANG LẦU 1
TL: 1/50

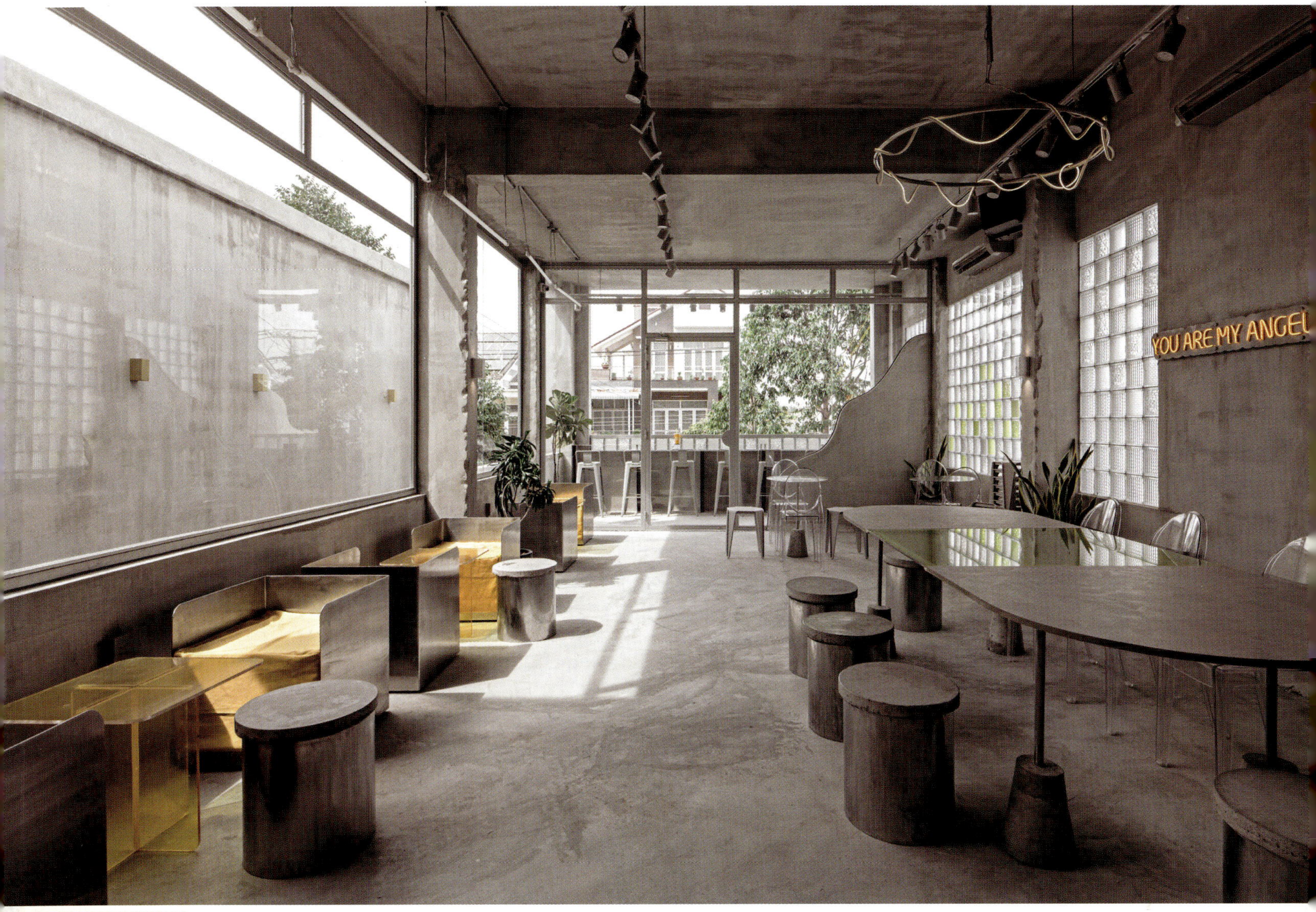

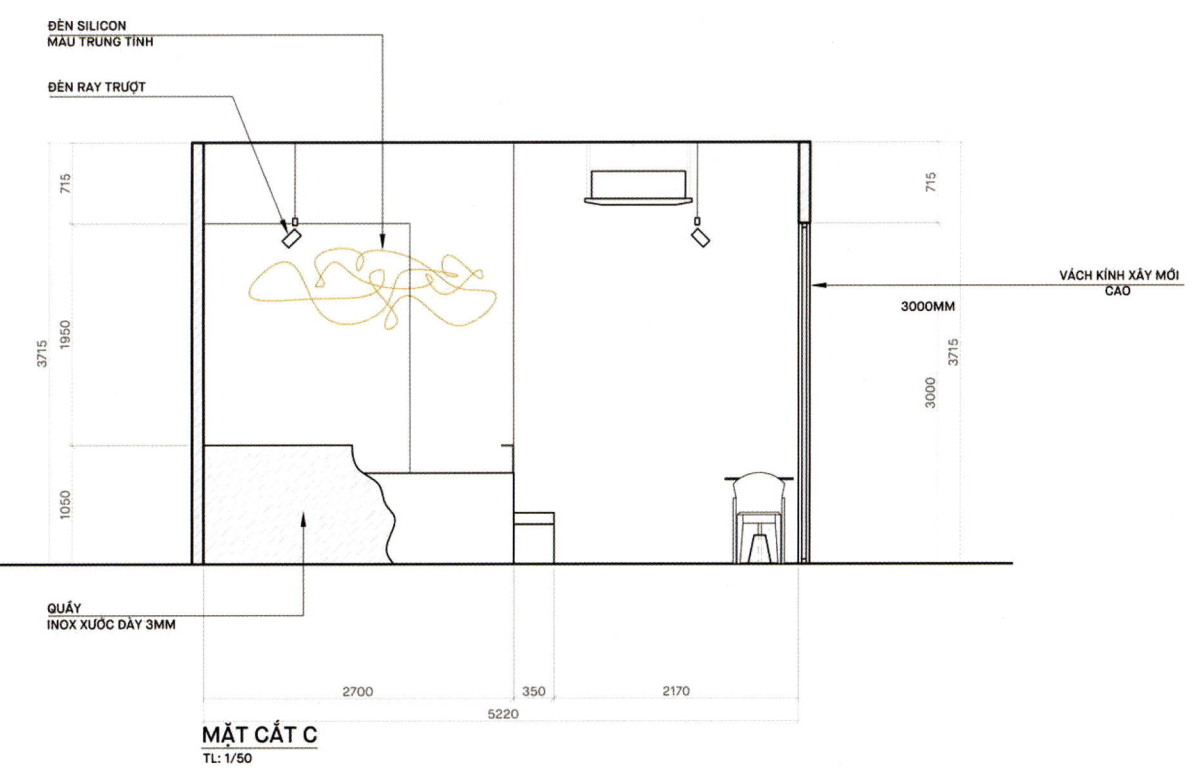

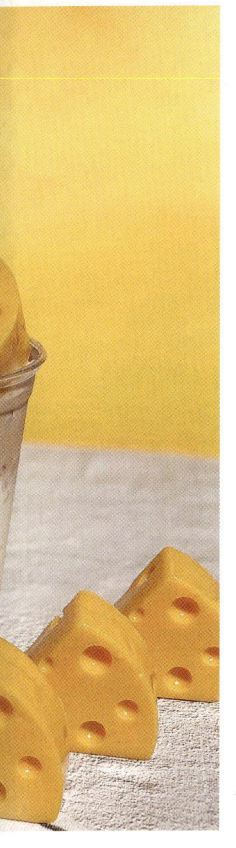

YOU ARE MY ANGEL

ĐÈN SILICON
MÀU TRUNG TÍNH
ĐÈN RAY TRƯỢT
VÁCH KÍNH XÂY MỚI
CAO
3000MM
QUẦY
INOX XƯỚC DÀY 3MM
715
1950
1050
3715
715
3000
3715
2700
350
2170
5220

MẶT CẮT C
TL: 1/50